DRUG DEVELOPMENT:
FROM LABORATORY TO CLINIC

DRUG DEVELOPMENT: FROM LABORATORY TO CLINIC

WALTER SNEADER

University of Strathclyde
Department of Pharmacy
Glasgow
Scotland

A Wiley Medical Publication

JOHN WILEY & SONS
Chichester · New York · Brisbane · Toronto · Singapore

Library of Congress Cataloging-in-Publication Data

Sneader, Walter.
    Drug development.

    (A Wiley medical publication)
    Bibliography: p.
    Includes index.
    1. Drugs—Testing.   2. Drug trade—Quality control.   I. Title.
    II. Series.   [DNLM: 1. Drug Evaluation.   2. Drug Industry—standards.
    3. Drug Screening. QV 771 S671d]
RS189.S64   1986   615'.1   86–3461

ISBN 0 471 91116 X (pbk.)

British Library Cataloguing in Publication Data:

Sneader, Walter
    Drug development: from laboratory to clinic.
    1. Drug—Research
    I. Title
    615'.191   RS122

ISBN 0 471 91116 X

Printed and bound in Great Britain

Contents

Preface

This book has been written to provide health care workers and others with a greater appreciation of the many skills required to overcome the diverse problems encountered by the pharmaceutical industry and clinicians seeking to develop safer medicines. Whilst it can be read either on its own or as a companion volume to my *Drug Discovery: The Evolution of Modern Medicines*, in no way can this text be considered a substitute for those specialist monographs which examine in considerable detail the topics that are only briefly outlined here.

The opening chapter discusses how advances in pharmacology and chemistry during the past century have meant that the principal modern medicines are synthetic organic compounds rather than highly poisonous drugs extracted from plants. This is followed by an account of how pharmaceutical research and development combines with an exemplary high standard of industrial production to enhance the quality and safety of medicines.

The greater safety of modern medicines often permits therapy to be continued for prolonged periods. Paradoxically, this has resulted in different hazards coming to the fore. Nowadays, adverse reactions to treatment are often of an allergic nature or due to chronic poisoning from repeated minor insults to vital organs. Such hazards are less predictable than those formerly arising from acute poisoning. Physicians must, therefore, be alert to the first signs of their appearance so as to be able to initiate remedial action before irreversible damage can occur.

Consideration is given to a variety of problems that have been encountered in the past. This should help the reader to appreciate the need for the extensive regulatory requirements that have become mandatory before a new medicinal product may be introduced into the clinic. Every conceivable effort is now made to identify any toxic effects that might arise after such a product is marketed. An outline of how post-marketing surveillance is achieved is accompanied by consideration of the inherent limitations. This leads on to a discussion of what can be done to identify those patients at greatest risk of experiencing adverse reactions.

Although drug therapy is now vastly more efficacious and safer than ever before, it retains a disturbing element of unpredictability. It is, therefore, inevitable that there is widespread interest in the activities of the pharmaceutical industry. The press and television are quick to report the occurrence of serious adverse reactions, sometimes in a highly critical manner. Whilst the ensuing criticism of the industry can be ill-informed, or the expectations of the public may be unrealistic, there have been occasions when media exposure has been of real benefit. The general public should be responsibly informed about the problems associated with the development and prescribing of new medicines. Hopefully, this short volume will help to do this.

Walter Sneader

Glasgow,
January 1986.

I

Drug Discovery

The thousand or so medicinal compounds in current use have been developed from a few score of prototypes, of which only a small proportion were able to meet all clinical requirements without the need for further development. These prototypes consist of active principles extracted from plants, mammalian hormones, antibiotics from microorganisms, and an assortment of synthetic compounds whose therapeutic potential was revealed by serendipity, a fortunate blend of chance and man's sagacity.

An impressive range of natural products was introduced for medicinal purposes by the Egyptians, Babylonians, Greeks and Romans. These substances, obtained from animal, vegetable and mineral sources, were often of an offensive nature as they had been selected with the intention of expunging demons thought to have taken possession of diseased bodies. Their effects on patients were quite often drastic since many were highly toxic. Even opium, the most valuable of all these ancient medicines, was not without its hazards. In the first century, Pliny the Elder warned of the dangers associated with its use. In this connection, it must be significant that the Greeks used the word *pharmakon* to mean either a poison or a drug. The implication is that they considered a drug to be a poison employed for medicinal purposes.

The introduction of the printing press in the fifteenth century led to the publication of herbals and pharmacopoeias. These became available to physicians and apothecaries throughout Europe, providing easy access to hitherto scarce information about ancient remedies. Consequently, the use of herbs and animal products as medicines progressed to reach its zenith during the seventeenth century. Thereafter, the excesses began to provoke a backlash from enlightened physicians who spurned the authoritarianism of the past in favour of the experimental method. During the following century the materia medica was gradually whittled down as many herbal and animal remedies were exposed either as worthless or as poisonous at the dose levels required to achieve an adequate therapeutic response.

PLANT POISONS

Before the advent of synthetic organic chemistry in the nineteenth century, plants were the main source of physiologically active substances that could be exploited for therapeutic purposes. In 1817, Friedrich Wilhelm Sertürner published his epochal report on how he and his companions had experienced the symptoms of severe opium poisoning after they had each swallowed 100 mg of a morphine salt. This substance, characterized as an alkaloid, was the first

1

active principle isolated from a plant as pure crystals. The subsequent extraction of more alkaloids, as well as glycosides, resulted in the availability of pure drugs with consistent quality and efficacy. Their enhanced reliability and safety were quickly exploited. Disappointingly, only a few proved satisfactory for clinical application and a mood of therapeutic nihilism began to spread. The hazards of morphine itself became all too apparent once the water-solubility of its salts was exploited with the development of the hypodermic syringe in 1853 by Alexander Wood of Edinburgh. Both the addictive liability and a tendency to kill by depressing the respiratory centre were increased when morphine was injected.

As the years have passed, plant poisons have lost their former attraction as a source of therapeutic innovation. Today, the *British Pharmacopoeia* lists about thirty active principles extracted from plants, of which only four have been introduced in the past half-century. Indeed, none of the world's leading pharmaceutical manufacturers now has a serious programme of research into drugs obtained from the plant kingdom. The reason for this is that plant products have all too often proved to be much more toxic than synthetic drugs. After all, alkaloids and glycosides are present in nature not to act as medicinal substances, but probably to protect plants by discouraging foraging predators. That is hardly an auspicious role for substances that are to be exploited for therapeutic purposes. It might be argued that plant poisons such as the vinca alkaloids or the semi-synthetic podophyllotoxin derivative etoposide have found medical application as cytotoxic agents in cancer chemotherapy—a rational extension of their role as toxins in nature. However, it must be pointed out that in 1955 the US Congress allocated millions of dollars to the Cancer Chemotherapy National Service Center, a part of the National Cancer Institute at Bethesda, for a massive screening programme. By the end of the decade, over one thousand compounds a month were being examined, many of which were extracted from plants. After more than twenty years of unrewarding screening, the programme to seek antitumour agents in plants was finally abandoned.

In order to appreciate the problems presented by plant poisons, it is instructive to consider some of those which still find therapeutic application, beginning with quinine. François Magendie promoted the use of this, the active principle of cinchona bark, by including it in his widely consulted *Formulaire* of 1821. Cinchona bark had been brought to Rome from South America in the sixteenth century, as a specific for malaria. Its popularity quickly spread throughout Europe, but such were its manifest dangers that religious bigotry conferred upon it the disparaging name of 'Jesuits' bark'. The inference behind this was that the bark was used to poison Protestants! It is said that Oliver Cromwell died rather than accept treatment with cinchona. The syndrome produced by excessive use of the bark, cinchonism, was characterized by abdominal pain, tinnitus, headache, nausea, rashes, and visual disturbances or even blindness. The occurrence of this became more frequent when John Brown of Edinburgh, who fancifully believed the intermittent fever of malaria was preceded by a state of debility, advocated the use of cinchona as a general tonic to stimulate the nervous system of asthenic patients. Since the Brunonian doctrine alleged that many diseased patients were asthenic, countless numbers of people were poisoned by misuse of cinchona as a tonic in desperate attempts to effect a cure. An alternative was to administer nux vomica as a tonic. It contained the convulsant poison strychnine. Misuse of plant poisons by supporters of Brown not only did much to hasten the demise of herbal medicine, but also stimulated the development of homeopathy by Samuel Hahneman of Leipzig in 1796. He administered infinitesimally small doses of plant poisons, which at least avoided intoxicating his patients. Early in the following century, the introduction of pure quinine salts reduced the incidence of cinchonism by ensuring that patients received a precisely determined dose of the alkaloid.

Leonhard Fuchs, the German botanist who gave the name *Digitalis purpurea* to the foxglove in 1542, considered this plant to be a violent poison. Yet, in 1785, William Withering published his celebrated book describing the beneficial effects of digitalis on patients with dropsy. He mistakenly believed the drug acted as a diuretic, failing to recognize that the diuresis was a consequence of the stimulant action on the failing heart. Despite this promising introduction into orthodox medicine, digitalis suffered much the same initial misuse as cinchona because it, too, was considered to be a valuable tonic. This is hardly surprising in view of its stimulating action on the heart, as well as its ability to induce both vomiting and purging. It was administered in the treatment of a diverse range of diseases presumed to arise as a consequence of asthenia, including insanity, epilepsy, depression, phthisis (pulmonary tuberculosis), pneumonia, diphtheria, typhoid, scarlet fever, measles, goitre, haemorrhoids and habitual abortion, not to mention all conceivable types of heart complaints. It was only in this present century that a clear understanding of the action of digitalis on the heart has emerged. Notwithstanding the introduction in the 1930s of the shorter-acting glycoside known as digoxin, many patients continued to be poisoned as a consequence of the often unrecognized cumulative toxicity. The availability of alternative therapies in recent years has resulted in a decline in the prescribing of this most hazardous drug.

Ergot was introduced into obstetric practice in the United States in the early part of the nineteenth century. European practitioners had shunned it because of its association with the outbreaks of ergotism resulting in tens of thousands of deaths in France and Germany during the Middle Ages. These were caused by the eating of bread prepared from rye infected with the ergot fungus. The American physician John Stearns claimed success after instituting the hitherto despised practice of German midwives who used either a powder or decoction of ergot to hasten prolonged labour. Others followed his lead, but often overlooked his warning that the *pulvis ad partum* should never be administered if there was reason to suspect that prolongation of labour might be due to an obstruction. So many still-births and maternal deaths ensued that one critic suggested the powder should be renamed *pulvis ad mortem*. To this day, ergot and its alkaloids have continued to be classified as potentially lethal drugs.

Considerable caution must always be exercised when prescribing ergotamine for migraine.

Ergot was not the only plant poison introduced by American physicians in the early years of the nineteenth century. They pioneered the use as a purgative of the juice expressed from the root of the May apple (also known as American mandrake), *Podophyllum peltatum*. Later, the more potent podophyllum resin gained widespread popularity throughout the world as a purgative, but its use for this purpose is now frowned upon since its action is produced through severe irritation of the intestinal mucosa.

Another poisonous plant that was not generally accepted in orthodox medical circles until the early years of the nineteenth century was the meadow saffron, *Colchicum autumnale*. Its history has many parallels with that of ergot, it being known to Dioscorides, a surgeon in Nero's army, as a highly poisonous plant. It was hailed as a remedy for gout by the Byzantine and Arabian physicians, and used by country folk in England. This was severely criticized by the medical profession, which described the plant as *Colchicum perniciosum*. However, the eminent London physician Sir Edward Home eventually established that it did indeed have some value in gout, despite its manifest toxicity. Shortly after, in 1820, Pelletier and Caventou at the École de Pharmacie in Paris isolated colchicine, which was somewhat safer since it was freed from the presence of veratrine, a powerful cardiac poison. Because colchicine frequently causes a variety of adverse reactions, it has been replaced by modern remedies for gout, such as allopurinol or indomethacin.

Several plants of the order *Solanaceae* have a long history of use as poisons. The best known of these is the deadly nightshade, the mydriatic action of which was exploited by the ladies of the Spanish court to beautify their eyes. This earned it the popular name of belladonna. Its notorious toxicity persuaded Linnaeus to confer upon it the botanical name of *Atropa belladonna*, Atropos being known to the Ancient Greeks as one of the three fates who could so readily cut the slender thread of life. The herb was eventually introduced into orthodox medical practice in the nineteenth century by ophthalmologists. They preferred to use atropine, the less irritant active principle, but soon discovered that toxic effects could be produced in the very young and the elderly by systemic absorption of even the small quantities instilled into the eye.

INORGANIC POISONS

The damaging effects of the plant poisons on patients were strongly criticized by the sixteenth century iatrochemist Theophrastus Bombastus von Hohenheim, also known as Paracelsus. During the inaugural lecture to mark his appointment to the chair of medicine at Basle he burnt the classic works of Galen, which had done so much to perpetuate the use of herbal remedies. Unfortunately, the metallic salts he recommended as an alternative were often just as harmful. His advocacy of the use of mercury preparations in the treatment of syphilis may have had some slight justification, but many thousands of patients consumed mercurials as 'alteratives' to alleviate a variety of other diseases, including meningitis, bronchitis, pleurisy, pneumonia, dysentery, rheumatism, hydrocephalus, dropsy and digestive disorders. The number of patients poisoned with mercury over the next four hundred years far exceeded that of those who benefited in any way. That mercury salts continued to be used until the middle of the twentieth century can be explained by the fact that their toxic effects often resembled the signs of the disease being treated. It was only in the early 1950s that a popular household remedy for teething infants (Steedman's Soothing Powders) was withdrawn from the British market after it had been discovered that the mercurous chloride in it caused pink disease.

Fowler's Solution, a preparation containing the potassium salt of arsenic acid, was introduced into the United Kingdom in 1786 by Thomas Fowler, a Stafford physician who had learned of the use of arsenic to treat malaria in Hungary. It became a popular tonic in the nineteenth century as a consequence of wildly exaggerated reports that the inhabitants of Styria in Austria kept themselves in good health by sprinkling their food with small amounts of arsenic. Use of arsenic continued until the 1940s when Fowler's Solution was still being prescribed as a tonic for patients with pernicious anaemia. This may have come about because of the heightened colouring of the cheeks arising from an increasing fragility of the blood capillaries consequent upon chronic arsenic poisoning. Amongst the many uses to which Fowler's Solution was put was that of treating sleeping sickness by David Livingstone, the Scottish medical missionary who explored much of Central Africa. The rational basis for this did not become evident until 1894, when Sir David Bruce demonstrated that trypanosomes could be temporarily eliminated from the blood by administration of Fowler's Solution. In an attempt to overcome the high toxicity of this and several organic arsenicals, which prevented a cure of sleeping sickness being achieved, Paul Ehrlich embarked upon his ambitious programme of research which led to the discovery of arsphenamine. This synthetic organic arsenical was tried successfully against syphilis in 1909 and, despite its somewhat toxic nature, was introduced into medicine the following year. It cured most patients and was only superseded when penicillin became available in the mid-1940s.

Another drug strongly favoured by Paracelsus was tartar emetic (antimony potassium tartrate), prepared by leaving white wine to stand in antimony vessels. He and his followers used this toxic compound with such reckless abandon that it was banned by the parliament of Paris in 1566. This is probably the earliest instance of statutory control over a medicinal substance. Notwithstanding the ban, a century later Louis XIV attributed his recovery from typhoid fever to the proscribed drug, as a result of which it not only regained its former popularity, but became even more widely recommended until its temporary demise in the late nineteenth century, again as a result of its persistent misuse. However, in 1906, Nicolle and Mesnil at the Pasteur Institute demonstrated that injections of tartar emetic could eliminate trypanosomes from the blood of laboratory mice, having much the same effect as Fowler's Solution. Field trials on infected cattle were disappointing because the effective dose produced too much toxicity. However, Gaspar de Oliveira Vianna, a young Brazilian physician, cured one of his patients suffering from a form of leishmaniasis by injecting tartar emetic. His first report of the treatment appeared in 1912, since when this highly toxic drug (sudden death after an injection was not unheard of) has saved many millions of lives by reducing the mortality rate from leishmaniasis to only 10 per cent, it having previously been around 90 per cent. Safer analogues have been developed for use in both leishmaniasis and schistosomiasis, but these still produce toxicity. Nevertheless, their widespread use emphasizes the point that the toxic properties of any drug must always be balanced against the possible benefits that may accrue from its use.

SYNTHETIC SUBSTITUTES FOR PLANT POISONS

The isolation of quinine and many other alkaloids hastened the development of pharmacology as an experimental science by ensuring a higher degree of reproducibility of results than had hitherto been attainable. The problem of biological variation between animals certainly remained, but rapid progress was made when pharmacology departments became established in universities during the second half of the nineteenth century. Parallel developments in the field of organic chemistry enabled chemists to consider the possibility of synthesizing substitutes for plant poisons. As progress occurred, attitudes towards the safety of drugs began to change. By the end of the century, synthetic drugs were starting to replace plant poisons.

The first drug for which synthetic substitutes were sought was quinine. The suggestion that it might be possible to prepare these as therapeutic agents was made by the professor of physiology at the University of Glasgow, John McKendrick. In 1875, he and the eminent chemist James Dewar had demonstrated that quinoline bases derived from decomposition of the alkaloid still retained some biological activity. Six years later, a tetrahydroquinoline base synthesized at the University of Munich by Wilhelm Koenigs and Otto Fischer was marketed under the name 'Kairin' after Wilhelm Filehne, professor of pharmacology at the University of Erlangen, had found it to be a promising antipyretic. Until then, quinine had been the most frequently prescribed drug for lowering temperature in febrile conditions, irrespective of whether due to malaria or not. In 1884, Filehne found phenazone ('Antipyrin') to be a far safer quinine substitute. This compound, mistakenly thought to be a quinoline derivative, had been synthesized by his colleague Ludwig Knorr. Like all the other quinine substitutes developed around this time, phenazone was devoid of antimalarial activity.

The commercial success of phenazone stimulated similar investigations into the preparation of analogues of morphine and cocaine. Both exhibited addictive properties which it was hoped would be absent in synthetic substitutes.

Today, cocaine is rarely used as a local anaesthetic, having been superseded by synthetic drugs, whilst much progress has been made towards finding non-addictive morphine substitutes. This process of improving upon nature has, on the one hand, also resulted in the development of valuable substitutes for salicin, atropine, ephedrine, papaverine, physostigmine, tubocurarine, dicoumarol, quinidine, theophylline, tetrahydrocannabinol and podophyllotoxin. On the other hand, chemists have so far been unable to develop truly superior substitutes for the digitalis glycosides, caffeine, eugenol, colchicine, ergometrine, pilocarpine, reserpine or the vinca alkaloids.

DRUGS FROM ANIMAL SOURCES

The administration of animal-derived products as medicines has a long history, dating back at least to Ancient Egypt. The foul concoction brewed by the witches in *Macbeth* faithfully depicted the state of the art, for the apothecary of Shakespeare's day required a veritable menagerie to supply crayfish, earthworms, frogs, foxes, lizards, scorpions, snails, swallows, toads, vipers and woodlice to compound powders, oils and syrups. Nor were their fat, bile, blood, urine, musk or excrement allowed to go to waste. As if this were not enough, assorted human organs, body fluids and waste products were also much in vogue. Thankfully, somewhat less bizarre animal products are now employed, although pregnant mare's urine was introduced in the 1930s as a major source of steroid hormones! The leech is still to be found in some dispensaries due to its value for stimulating blood flow around skin grafts.

The modern use of glandular extracts traces its origins to the notorious experiment by an eminent Parisian physiologist Charles-Édouard Brown-Séquard who, in 1889 at the age of 72, reported that he had been rejuvenated by injecting himself with extracts of dog and guinea-pig testicles. The opportunistic surge of interest in organotherapy that ensued was not entirely devoid of merit, for it soon resulted in the introduction of thyroid and adrenal medullary extracts. From the latter, adrenaline (epinephrine) was isolated as the first pure hormone in 1901, its synthesis being achieved four years later. Pure, crystalline thyroxine was not isolated until 1914, by which time extracts of the pituitary gland were being used to inhibit excessive urine production in patients with diabetes insipidus. By then, pancreatic extracts had also been administered to a few diabetic patients, but it was not until 1921 that Banting and Best isolated insulin. This proved to be a turning point in drug research, with the lives of millions of diabetics being saved with insulin injections. The resulting stimulus given to research into pituitary, adrenal cortex and sex hormones over the ensuing two decades cannot be overlooked.

The success of modern hormone therapy emphasizes why plant poisons have been abandoned as a prime source of therapeutic agents. The natural role of hormones is to modify physiological function, not to disrupt it. Nevertheless, it has required considerable scientific ingenuity to make such therapy a reality, for hormones have often been difficult to isolate in pure form as they tend to be unstable. Hormones such as adrenaline and hydrocortisone have lacked sufficient specificity of action, but the quite remarkable efforts of medicinal chemists have overcome even this drawback.

RATIONAL APPROACHES TO DRUG DISCOVERY

In the early 1890s, Paul Ehrlich observed that the toxicity of cocaine substitutes to the liver did not necessarily parallel their potency as local anaesthetics. Several years later, he turned this observation to advantage in his quest for antiprotozoal agents safer than the inappropriately named 'Atoxyl', an arsenic-containing aniline derivative. By assaying both its efficacy in curing infected mice and its toxicity to these animals, he derived a therapeutic index for each compound as a quantitative measure of its medicinal potential. After more than six hundred closely related arsenicals had been examined, Ehrlich and Hata established that arsphenamine could cure mice experimentally infected with syphilis. Clinical trials then confirmed that although still somewhat toxic, the new drug could be used with caution in humans. It was the first truly effective synthetic chemotherapeutic agent.

At the heart of Ehrlich's approach was the concept of selectivity of drug action. As he applied this in the sphere of chemotherapy, his ultimate objective was to find a compound that he described as parasitotropic without being organotropic. Thus, arsphenamine was a success because it did not damage the optic or vestibular nerves as earlier arsenicals had done, yet it was highly effective in killing spirochaetes. Ehrlich would never have embarked on the long process of screening so many arsenicals had he not been in possession of the evidence that a prototype compound, 'Atoxyl', already exhibited some of the desired selectivity. Indeed, this remains the *sine qua non* which defines a drug prototype; without evidence of a degree of selectivity of action in the lead compound, any attempt to synthesize useful analogues degenerates into a random effort that is unlikely to succeed. Skill in identifying novel prototypes with some desirable selectivity of action, coupled with the ability to devise a reliable biological screening procedure, remains the key to success in drug research.

The approach devised by Ehrlich still underpins much contemporary drug research. It was based on a series of small, stepwise advances towards the ultimate goal. As new compounds were tested in laboratory animals, and in one or two cases even in patients, Ehrlich received valuable feedback of information relating to the influence of structural modification on biological activity. To describe this procedure as 'molecular roulette' is to disparage the most practical method through which the pace of drug discovery can be accelerated. It is, in fact, difficult to cite an example of a *de novo* approach whereby a drug prototype has been rationally designed without being modelled on an existing drug, natural product or hormone. Possibly, it could be argued that Peters's development of dimercaprol (British anti-Lewisite) during the Second World War comes into this category. His work was based on the postulated mode of interaction between the arsenical chemical warfare agent Lewisite and thiol groups in the vital pyruvate oxidase enzyme. Dimercaprol was fabricated as a sacrificial molecule containing two adjacent thiol groups to interact preferentially with Lewisite. This kind of approach to drug synthesis is rare, perhaps because all too little progress has been made in understanding how chemical structure influences biological activity. There is still no general theoretical framework that enables medicinal chemists to anticipate much about either the pharmacological or the toxicological properties of novel compounds.

Ideas advanced in the earlier part of this century still serve as guidelines for chemists seeking to design novel drugs. In 1905, after demonstrating that nicotine stimulated muscles to which the nerve supply had been cut, and that this stimulation could be antagonized by curare, John Langley promulgated his theory of 'receptive substances' to explain both this and the antagonism he had previously detected between pilocarpine and atropine. He believed that in each case the receptive substance could accept the stimulating compound or be blocked by the antagonist. Paul Ehrlich refined the concept, seeing a parallel between tissue receptors for drugs and the active sites of enzymes into which natural substrates could fit. This has become the central dogma of drug research, receptors being implicated in a variety of processes that involve their binding with ligands that may be either their normal substrates or drugs. These processes include carrier systems which transport ligands across membrane barriers, storage systems that retain the ligand until its release is required, and the triggering of sequences of events associated with hormonal activitity. The latter has proved extremely important for the development of new drugs, be they compounds which mimic or which antagonize the action of hormones.

Industrial pharmacologists devise sophisticated screening procedures which can identify drugs with enhanced selectivity of action. Whenever any promising compound emerges from such screening, chemists submit closely related analogues. These will often have been designed by a medicinal chemist who has taken considerable care to select only those compounds that could pinpoint chemical features or physical properties responsible for any further change in selectivity of action. The process of screening and synthesis usually has to be repeated many times before a promising compound emerges. Projects often have to be abandoned. An example of a successful outcome, however, is afforded by the modification of the structure of orciprenaline. This was itself a selective analogue of adrenaline that retained bronchodilator activity at beta-adrenoceptors without producing the pressor response mediated by alpha-adrenoceptors. Screening of an orciprenaline analogue with an extra methyl group attached to its nitrogen atom substituent revealed it to be more or less free of cardiac side-effects. This was because this new analogue, terbutaline, was too bulky to fit the beta-adrenoceptors in the heart. Several other hormone analogues and antagonists have been designed to affect selectively certain hormonal properties in order to obtain therapeutic advantages. For example the introduction of a 16-hydroxyl group reduced the salt-retaining (mineralocorticoid) activity of adrenocortical steroids without affecting their anti-inflammatory activity. Again, steric hindrance of fit to one of two different types of hormone receptor was involved. Selectivity can also be increased by specifically enhancing the fit to one type of receptor. This happened when hydrocortisone was converted by mould fermentation into prednisolone. The salt-retention activity of the new steroid was similar to that of hydrocortisone, but a fivefold increase in its anti-inflammatory activity permitted less troublesome therapy. Several highly potent corticosteroids with greatly enhanced anti-inflammatory activity, coupled with reduced salt-retention activity, were subsequently prepared by the incorporation of a 16-hydroxyl function, e.g. triamcinolone and dexamethasone.

Medicinal chemists consider the binding of a drug to an enzyme as yet another example of drug–receptor interaction. The importance of drugs which inhibit those enzymes that exert a key role in certain complex physiological processes has long been recognized. In 1926, Loewi demonstrated that physostigmine caused accumulation of acetylcholine by inhibiting acetylcholinesterase. A recent example of a novel application for an enzyme inhibitor relates to the angiotensin-converting enzyme, and was based on the discovery of the ability of a small peptide from a snake venom to inhibit this enzyme which catalyses the conversion of angiotensin I into angiotensin II, the most potent hypertensive substance in the body. Non-peptide analogues of the peptide were examined as potential inhibitors of the angiotensin-converting enzyme, the influences of small changes in molecular geometry being carefully charted. This resulted in the development of captopril, the first orally active member of a promising new class of antihypertensive agents.

The design of enzyme inhibitors is frequently based on the antimetabolite concept. This requires the synthesis of molecules closely resembling metabolites which are the normal substrates in enzymic reactions that subserve vital roles in the living cell. The antimetabolite concept was elaborated in the early 1940s in the wake of the introduction of the antibacterial sulphonamides. At that time, the 'lock and key' analogy was used to describe sulphanilamide as a false 'key' which blocked the entry of 4-aminobenzoic acid, the true 'key', into a 'lock', *viz*, the active site of an enzyme present in bacteria. Antimetabolites have been of considerable value in cancer chemotherapy, where methotrexate, mercaptopurine, fluorouracil and cytarabine are capable of directly or indirectly interfering with enzymes responsible for the elaboration of DNA prior to cell replication. Although the antimetabolite approach still offers some prospects for rational drug discovery, many potential antimetabolites have been found to lack any useful therapeutic activity. As time has passed, the rate of introduction of clinically effective antimetabolites has declined. It is, however, probable that the future will see the further development of selective enzyme inhibitors based on the slight differences between enzymes common to different organisms. This is typified by the antibacterial agent trimethoprim, which inhibits bacterial dihydrofolate reductase some fifty thousand times as effectively as it does mammalian enzyme. The outcome is that DNA synthesis is disrupted in pathogenic organisms.

Drugs not only inhibit enzymes but can themselves be substrates for them. The brief duration of action of both established and potential new drugs may be due to rapid metabolic degradation. Analogues which cannot act as substrates for the destructive enzymes have been prepared when longer-acting drugs are required. For example, introduction of a methyl group adjacent to the sensitive ester function of acetylcholine, to form methacholine, blocked the fit of the hormone to acetylcholinesterase and overcame the problem of its transient action. Possibly the most successful example of this use of steric hindrance to enzymic destruction is the placing of bulky substituents on the side chain of semi-synthetic penicillins such as methicillin and cloxacillin to confer resistance to penicillinase produced by staphylococci. Medicinal chemists have exploited the introduction of a methyl or halogen at the 6-position of the steroid nucleus to block oxidative metabolism. An alternative way of avoiding rapid metabolism of drugs involves replacing a metabolically sensitive function by a more robust one. This has been achieved with the carbamate ester analogue of acetylcholine known as carbachol, and with the antidiabetic agent chlorpropamide in which a chlorine atom replaces the oxidisable methyl group of tolbutamide.

In the past 20 years, the process of drug design has also taken into account the influence of chemical structure on drug transport through the body. Medicinal chemists have acquired a clearer understanding of how physical properties can influence the transport of drugs across membranes. The onset, intensity, duration and quality of drug action are now seen to depend on the ease with which the drug can cross a variety of membrane barriers. This gives scope for introducing changes in the molecular structure of existing drugs to alter their physical properties in such a way as to influence their absorption, distribution, metabolism and excretion. For example, lack of oral activity may be due to the inability of a molecule to diffuse through the gut wall because of inadequate lipid solubility, whereas side-effects on the central nervous system may be caused by excessive lipid solubility. This same physical parameter may also influence ability to enter liver cells, thereby affecting the rate at which the drug is metabolized. Whether a drug is reabsorbed back into the circulation via the kidney tubules can depend not only on lipophilicity, but also on its tendency to ionize, as indicated by its dissociation constant. The accumulation of molecules in fatty tissue is similarly affected, as is penetration through the skin, into the eye and so forth.

Organic chemists are able to synthesize an unlimited range of analogues of any compounds with interesting biological activity. Regrettably, the literature of medicinal chemistry contains many reports of novel compounds that give the impression of having been synthesized principally to avoid coming within the scope of existing patents or else to further the career of those who must publish or be damned. In the past, manufacturers sometimes developed analogues of a rival company's latest product in order to capture a share of a lucrative market. This happens less frequently nowadays because of the risk and costs involved in proceeding through the lengthy process of safety testing and clinical trials in order to satisfy drug regulatory authorities. Commercial considerations aside, it has to be admitted that the possibility nevertheless exists that a 'me too' drug of this type might offer some unanticipated advantage, such as a reduced incidence of a troublesome side-effect. However, the equal possibility of an unanticipated serious adverse reaction has to be set against this. The issue is far more problematical than some critics of the pharmaceutical industry would have us believe. If the true advantage of a new drug is that it exhibits a reduced incidence of a life-threatening adverse reaction which affects less than one in a thousand patients, this cannot be revealed even by comparative clinical trials on two or three thousand patients. Only by making such a drug freely available can it be used on sufficient numbers of patients to prove its greater safety. Nor is this merely a hypothetical situation, for most modern drugs produce life-threatening adverse reactions in less than one in twenty thousand patients.

The sensible prescriber always demands good reason for prescribing a new product. He will never accept that chemical novelty alone is sufficient justification for this. If this is ever claimed as the sole advantage of a new product, he will have grounds for suspecting that patent circumvention is the underlying motive behind its introduction. After all, thalidomide was promoted mainly on the grounds that it was 'a non-barbiturate hypnotic'.

EXPERIMENTS ON ANIMALS

The success of the Ehrlich approach to drug discovery has depended upon the use of live animals to identify drugs with enhanced selectivity of action. The search for alternatives to live animal experiments is vigorously pursued for a variety of reasons, not least of which is the expense involved in maintaining animals in decent conditions. Many of the biological tests required for the detection of compounds with improved selectivity of action can be conducted on tissues isolated from animals that have been killed, or occasionally even on tissue culture preparations. This cannot be done where the intention is to improve the transport and distribution of a drug in order to enhance its pharmacokinetic profile.

Those who oppose the use of animals in drug research do so because they believe that animals can suffer as much as humans; their dilemma is that the total abandonment of animal experimentation would deny all hope of therapeutic advance to relieve the distress of both human and animal victims of diseases for which no effective medicines yet exist. Those who instead call for the reduction in the number of animal experiments to the minimum necessary to develop and ensure the safety of essential new drugs have the support of most researchers.

SERENDIPITY IN DRUG RESEARCH

Rational approaches to drug design still depend on the availability of suitable prototypes which can be structurally modified to produce more selective compounds. The most fruitful source of these prototypes has been that combination of chance and sagacity which Horace Walpole described as 'serendipity'. He coined this term more than two hundred years ago after reading a poem about the three princes of Serendip (Sri Lanka) who repeatedly made discoveries they were not seeking.

Serendipity lay behind the introduction of both ether and nitrous oxide anaesthesia in the 1840s. The discovery of their ability to nullify awareness of pain caused by accidental injury was recognized through their inhalation for recreational purposes. In contrast, the anaesthetic properties of ethylene and of cyclopropane were revealed when small animals were exposed to these gases in order to confirm expected toxicity! It has to be admitted that not very much sagacity was required to recognize anaesthetic or hypnotic compounds. Serendipity was responsible for the discovery of a diverse range of these, including methaqualone, chlormethiazole and the respective prototypes from which propanidid, etomidate and several steroids were developed as intravenous anaesthetics. To this list of compounds that depress the activity of the brain must also be added several anticonvulsants, including the bromides, phenobarbitone, troxidone, the benzamide precursor of beclamide, carbamazepine and valproic acid.

The complexity of the biochemistry of the brain has meant that it has not been possible to design effective psychopharmacological agents. Those that are used clinically have been based on serendipitous discoveries. Chlorpromazine was synthesized for use in the prevention of shock during surgery, but happened to induce an unusual calming effect on patients. Another major tranquillizer, haloperidol, was developed after an analogue of pethidine was found to have more than a straightforward analgesic action on laboratory animals. The antidepressant drugs imipramine, iproniazid (the first monoamine oxidase inhibitor) and mianserin also owe their clinical role to astute exploitation of their unanticipated mood-elevating properties, which were first observed in patients. The serendipitous discovery of the muscle-relaxant activity of glycerol ethers designed to be used as antibacterial agents was made during their toxicological testing when a flaccid paralysis of the limbs of mice was noted. This resulted in the development of mephenesin and the anti-anxiety agent meprobamate. The relaxants chlormezanone and dantrolene were also discovered serendipitously.

The profound effects of the psychotomimetic agent lysergic acid diethylamide (LSD) were revealed when it caused vivid hallucinations after the chemist preparing it for study as an oxytocic accidentally ingested a trace of the compound. Although morphine remains the drug of choice for the control of severe pain, the serendipitous discovery of the analgesic properties of pethidine (meperidine) cannot be overlooked. It was synthesized as an atropine analogue for use as an antispasmodic agent. During testing on cats it caused their tails to flip backwards in a manner previously observed with narcotic drugs. This led to its introduction as a morphine substitute.

The widespread use of the clinical thermometer in the latter third of the nineteenth century ensured that any drug capable of lowering body temperature in feverish patients would be detected. The discovery of the antipyretic properties of acetanilide was due to its being wrongly supplied instead of naphthalene. When given to patients with intestinal worms, its antipyretic properties were revealed. Phenacetin and paracetamol were ultimately derived from it. The antipyretic activity of salicylic acid, an analogue of phenol (carbolic acid), was revealed when it was tried as an internal antiseptic in typhoid patients. Later, it was administered as an antipyretic to relieve rheumatic fever, only to reveal that it also had anti-inflammatory properties!

In the field of local anaesthesia, the unanticipated properties of cyclic derivatives of acetanilide, synthesized as potential antipyretics, were fully exploited with the development of cinchocaine (dibucaine). The same happened with isogramine, from which lignocaine was developed, after the chemist who first synthesized the alkaloid had noticed that it numbed his tongue.

Several drugs acting on the autonomic nervous system have been discovered serendipitously, such as the imidazolines, dibenamine and dichloroisoprenaline (the forerunner of the beta-adrenoceptor antagonists), all of which affect adrenergic transmission. The organophosphorus anticholinesterases and various antihypertensive agents were also discovered serendipitously. These include hexamethonium, mecamylamine, xylocholine (the lead from which bethanidine was developed), the forerunner of guanethidine, and methyldopa.

It is not really surprising that diuretics have been discovered serendipitously. The first truly potent diuretic, the mercurial anti-infective agent merbaphen, was discovered by nursing staff to produce an intense diuresis when administered by injection to a syphilitic patient. Some years later, the systematic exploitation of the alkaline diuresis following the administration of sulphanilamide led to the development of acetazolamide and the thiazides. The thiazide analogue diazoxide, however, turned out to be devoid of diuretic activity, yet it still proved of value as a hypotensive drug. It was not the only cardiovascular agent to exhibit a different mode of action from that which had been intended. Much the same occurred with amiodarone which was prepared as a potential coronary dilator but turned out to be an antiarrhythmic agent. The similar activity of flecainide was observed serendipitously as a consequence of screening a series of fluorine-substituted benzamides.

Serendipitous discoveries can be highly profitable, a noteworthy example being the exploitation of a patient's comment that her travel sickness had disappeared whilst her allergy was being treated with dimenhydrinate, a salt of diphenhydramine. Commercial success also followed the exploitation of the observation that some acids amongst a series of khellin analogues being tested as bronchial relaxants actually owed their protective action against artificially induced asthma to their ability to stabilize mast cells. This resulted in the development of sodium cromoglycate (cromolyn).

The field of hormone research is one area in which rationality has provided most of the new drugs. Nevertheless, it should be noted that the discovery of the antithyroid drugs arose from an investigation into the influence of sulphaguanidine on the production of vitamins by bacteria in the gut. Investigations into the toxicity of another antibacterial sulphonamide, carbutamide, resulted in recognition of its hypoglycaemic activity and led to the development of tolbutamide and other oral antidiabetic agents. Several important steroids were introduced in the wake of the discovery of the unexpected oral activity of the 17-acetoxy ester of progesterone, originally prepared as a long-acting derivative for depot injection.

The development of antibacterial chemotherapy has depended entirely on the serendipitous discovery of prototypes. The first step occurred when cultures of trypanosomes and pneumococci were accidentally confused in a crowded hospital laboratory, resulting in the recognition of certain biochemical similarities between them. Consequently, the pneumococci were exposed to antitrypanosomal agents, some of which were active against them, including ethylhydrocupreine. However, the most celebrated example of serendipity in drug discovery must unquestionably be Fleming's astute observation of the antibiotic activity of a *Penicillium* mould. Admittedly, this was far from being the first observation of this phenomenon, but it paved the way for Florey and Chain to develop penicillin.

No field of drug research has been as strongly influenced by serendipitous discoveries as has cancer chemotherapy. The observation that castration of dogs being prepared for studies on the effects of testosterone resulted in the regression of spontaneous prostatic tumours was rapidly exploited when steroid hormones were introduced with considerable success in a variety of hormone-dependent malignancies. Of course, not all drugs used in such cancers are hormone analogues. The revelation that the chronic toxicity of the anticonvulsant aminoglutethimide was due to its inhibition of steroid synthesis has been put to good effect in dealing with breast cancer.

The unexpected discovery that the systemic toxicity of the nitrogen mustard chemical warfare agents was due to damage to rapidly proliferating cells was immediately turned to advantage in the control of several types of tumours. Nevertheless, the most far-reaching of the serendipitous discoveries in cancer chemotherapy occurred when Farber found that folic acid conjugates, far from relieving leukaemia, actually accelerated the fatal progress of the disease. Within months, he went on to demonstrate the value of antifolate therapy in this disease, thereby preparing the ground for the widespread use of antimetabolites. Also of value in leukaemia is vincristine, one of the alkaloids from the vinca plant which caused a reduction in the number of circulating white blood cells in animals being studied to assess its reputed antidiabetic action. Another instance of serendipity in cancer research was the recognition that exposure of bacterial cells to an electric current caused the formation of a platinum complex which interfered with cell replication. This led to the ultimate development of cisplatin therapy. Finally, mention must be made of the useful activity of a precursor of razoxane, a by-product from the synthesis of a series of metal-chelating agents under investigation as potential cancer chemotherapeutic agents. Although razoxane was unable to chelate metals, it turned out to have therapeutic value that was lacking in the compounds that were chelating agents.

The financial rewards obtained from products that originated from screening programmes designed to find more selective analogues of existing drugs may not, in general, match those that can sometimes result from the exploitation of serendipity. However, the small number of serendipitous discoveries over the years does not inspire confidence in the future of any drug company that seeks salvation solely from this source. Consistent success has been achieved by a small number of companies which have invested in extensive screening programmes in attempts to force the pace of innovation. However, there are also improvements that can be achieved other than those relating to the biological activity of drugs. These relate to the chemical and pharmaceutical properties which affect the formulation of drugs. For instance, either a lack of adequate water solubility or chemical instability may make it impossible to formulate a drug as an injection or liquid dosage form. An obvious case in point is that of aspirin, where its sensitivity towards aqueous hydrolysis has yet to be overcome in an acceptable liquid formulation.

II

Production and Formulation

When the medicinal chemist synthesizes compounds for screening, he seeks the most convenient route that will enable a reasonable number of compounds to be prepared quickly. The chosen route will not have been designed for large-scale economic production; a different approach is required for this. It will be necessary to build a pilot plant not only to supply material for formulation and clinical studies, but also to allow chemical engineers to assess economic and production aspects of manufacture prior to drawing up their plans for the full-scale plant. Much effort will be put into improving the yield and eliminating impurities formed from side-reactions at every stage of the chemical synthesis. By the time these steps have been taken, there will be considerably more information available indicating whether the drug is worthy of further development. All too often, the project will be abandoned at this point since development costs from here onwards will become so great that only outstanding compounds can justify the capital risk involved. Herein lies

Figure 1.   The laboratory in John Bell's Pharmacy at 225 Oxford Street, London, in 1840. From an engraving by J.C. Murray after W. Hunt. *Reproduced by permission of the Wellcome Foundation Library, London.*

17

18

Figure 2. Modern pilot plant area. *Reproduced by permission of Glaxo Group of Companies.*

the cause of the resentment felt by the leading research-based pharmaceutical companies when governments compel them to issue licences to rival manufacturers who have not taken this risk. The loss in sales revenue incurred in certain countries forces up the price paid elsewhere so that manufacturers may recoup the hundred million pounds or so that will be required to develop a major new drug. International agreements could put an end to what amounts to government-sponsored piracy.

The patent life of a new drug is 20 years, but nowadays it frequently requires 10–14 years of development work and clinical testing before permission to market a drug is granted by regulatory authorities. It is, therefore, essential that the plans for the large-scale production of any promising compound be considered at the earliest possible moment so as to avoid delay if the preclinical evaluation ultimately proves satisfactory. Commissioning chemical production plant is always an expensive and protracted stage in the development of any new drug. Management faces a critical decision in assessing if and when to initiate this stage of the project in the light of the accumulated pharmacological, toxicological and clinical information. The intention is to hold back from the considerable expenditure that could be wasted if the project has to be abandoned, whilst ensuring that the chemical plant will be functioning smoothly as soon as permission to market the product has been received. To help

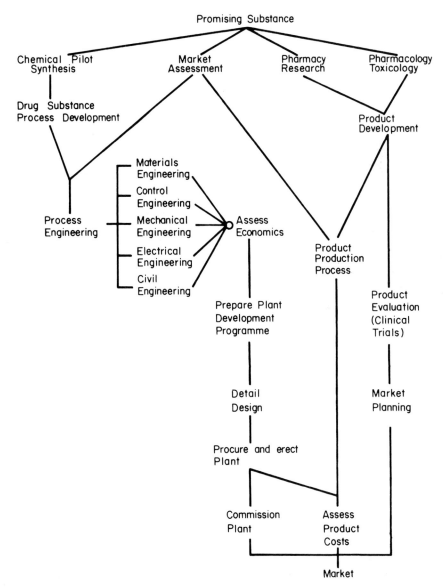

Figure 3. Plant process design plan for a new drug. *Reproduced by permission of The Pharmaceutical Journal and K.A. Lees.*

management make this difficult decision, it will be provided with data obtained from experience of the pilot-scale production of the material used in formulation development and early clinical studies.

The nature of the production problems arising with any new drug depends on a variety of factors. The scale of the production is influenced by marketing considerations and by the nature and size of dosage forms, as well as their frequency of administration to patients. Close collaboration between engineer and formulation pharmacists must be maintained. Skilful work by the latter can considerably increase the efficiency with which drugs can enter the body minimizing the amounts required, with a consequent reduction in the size of chemical plant that has to be fabricated.

PREFORMULATION STUDIES

Consideration must be given to problems that arise from the scaling-up of production that will be required when a new drug is eventually marketed. For example, handling an inflammable solvent presents no hazard in the laboratory, but the situation changes when large plant is involved. It is also one thing to prepare a batch of ten thousand tablets on a small machine and quite another to turn out a million tablets in an afternoon on a piece of high-speed equipment that requires a constant flow of powder of guaranteed consistency. Since any variation in the physical properties of the ingredients can then be disastrous strict control over their quality is essential. The modern approach is to have quality assurance built into the entire production process from the preparation of the raw ingredients onwards.

The concept of building quality assurance into the process from the outset can be illustrated by the preformulation studies that were carried out on a non-steroidal anti-inflammatory drug known as fenoprofen, which is a derivative of 2-phenylpropionic acid.

Fenoprofen as the free acid was unsuitable for solid dosage forms since it melted just above room temperature when pure, and normally existed as a viscous oil. The obvious way round this problem was to prepare a high-melting salt by treating the acid with an aqueous alkaline solution. The first requirement was for a crystalline salt as this would be of high purity and easier to process. The potassium salt turned out to be very hygroscopic, which means that a tablet containing it would be difficult to manufacture and could swell and disintegrate as water was soaked up. The magnesium salt did not crystallize at all, whilst the aluminium salt was amorphous and rather insoluble in water. This latter property ruled out its use in tablets or capsules, but left open the possibility of its use in an oral suspension. It proved impossible to obtain anhydrous crystalline sodium or calcium salts, but their dihydrates were readily isolated as pure crystals. Although the sodium salt was much more water soluble than the calcium salt, it had an unacceptable tendency to dehydrate at room temperature under conditions of low humidity. The calcium salt remained stable as its dihydrate at temperatures up to 70 °C, thus ensuring it suitability during the manufacturing processes and in formulations to be marketed in countries with hot climates. The relatively weak attachment of the water of crystallization in the sodium salt also permitted it to undergo chemical reaction with salts of organic bases such as dextropropoxyphene or codeine, which were being considered as ingredients of compound tablets containing fenoprofen. The greater stability of the calcium salt dihydrate ensured there would be no release of any water of crystallization that could permit the bases to enter solution and react with the acidic fenoprofen.

The fact that the dihydrate of the calcium salt of fenoprofen had ideal physical properties did not in itself guarantee it would have adequate bioavailability. Some drugs can exist in a variety of different crystalline forms, or habits, that are known as polymorphs. The ability of different polymorphs to dissolve in the gastro-intestinal tract can vary somewhat, and in extreme cases certain polymorphs are devoid of biological activity. For example, chloramphenicol palmitate exists as two stable polymorphs, only one of which has antibiotic activity when administered orally. Since this can be a life-saving drug in the treatment of typhoid, great caution must be exercised to ensure that any formulation of this antibiotic contains the active polymorph. Fortunately, the presence of the inactive polymorph can be detected through a slight difference in its ability to absorb infrared light at certain wavelengths. Inactive polymorphs of phenobarbitone and some sulphonamides are also known. When tests on fenoprofen calcium dihydrate showed there were no inactive polymorphs and that it was well absorbed from the gut, it was finally accepted as the appropriate salt for all solid dosage formulations of the drug.

The oral bioavailability of a drug with poor water-solubility can be markedly affected by the size (or, more precisely, the size distribution) of the crystalline particles of it in the chosen formulation. If the particles are too large, there may not be enough time for them to dissolve completely during transit through the upper reaches of the small intestine where absorption occurs. Following a change in the manufacturing process of a leading brand of digoxin tablets in 1969, it was eventually discovered that plasma levels were halved. This was caused by a reduced dissolution rate attributable to a change in the particle size.

The development of advanced milling techniques for both dry and wet grinding has permitted extremely fine powders to be produced on a large scale. By micronizing crystals in this way, their overall surface area is so greatly increased that their dissolution rate in the gastro-intestinal tract is significantly enhanced. This approach was used successfully some years ago to improve the absorption of the antifungal antibiotic griseofulvin. Reduction of particle size can also assist penetration of some drugs administered to the lungs by aerosol. Additional advantages are gained by reducing the particle size of oral suspensions of insoluble drugs, whereupon the particles become lighter in weight and so remain dispersed longer. Particle size reduction can also improve the free-flowing characteristics of bulk powders fed rapidly through hoppers into encapsulating and tabletting machines. However, the shape of the fine crystals is also important in this aspect as well as in influencing the bulk density of the powder, and hence the size of the tablets. Sometimes it is possible to obtain a smaller tablet by altering the conditions under which a drug is crystallized, thereby avoiding the formation of either flat plates or long needles which increase the bulk of the raw product.

PHARMACOKINETIC CONSIDERATIONS

Pharmaceutical companies initiate pharmacokinetic studies of a new drug in volunteers as soon as it can be considered safe to do so. These are designed to quantify the absorption, distribution, metabolism and excretion of the drug. This affords an indication of whether there is scope for improving safety or efficacy by using the skills of the formulation pharmacist. It may be possible to compensate for failure to maintain optimal blood levels attained by the drug during the intended period of treatment. Failure to face up to this type of situation can result in the rejection of an otherwise promising new drug.

The attainment of adequate blood levels has to be carefully considered since

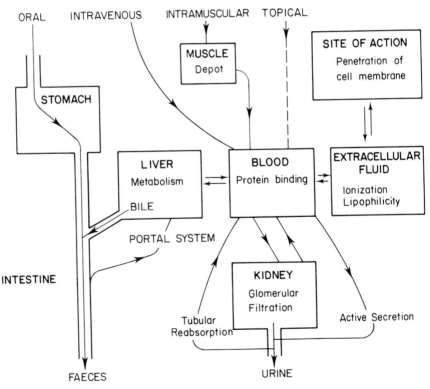

Figure 4. Diagrammatic representation of the disposition and fate of a drug.

the concentration of a drug in the plasma is usually a reasonable predictor for both efficacy and toxicity. If, on the one hand, too much drug enters the circulation soon after a dose has been taken, the high plasma levels might lead to unpleasant toxic effects. On the other hand, if plasma levels fall too quickly, the therapeutic action might only be transient. Clearly, the most desirable profile will depend on the nature of the drug and the desired therapeutic action. For instance, the plasma levels of a bacteriostatic sulphonamide must be maintained above the minimum concentration required to inhibit bacterial replication for long enough to enable the immune system to eradicate the infection. If this level cannot be maintained between doses, the infection may quickly run out of control. Increasing the dose significantly above the minimum necessary to maintain the desired bacteriostatic level may exceed the threshold at which toxic effects begin to occur.

Formulation pharmacists speak of the 'therapeutic window' to describe the range within which the plasma level of a drug should be maintained to ensure an optimal therapeutic response. For many existing compounds, this window is so wide that little skill is required to keep within it. A hard gelatin capsule, or a starch-based tablet which disintegrates on being moistened in the stomach, will suffice in such cases. However, with the greater safety now demanded of new drugs, there is often little room for manoeuvre.

SELECTION OF DOSAGE FORMS

The formulation pharmacist acts under a variety of constraints when deciding which dosage forms to develop. Marketing requirements may have excluded certain approaches. For example, there would be reluctance to introduce a parenteral injection of a drug intended for chronic use, but a life-saving antibiotic could be presented in this form with equanimity. Similarly, a drug for infants would be required as concentrated drops rather than as tablets. Marketing factors aside, some drugs cannot be given as oral formulations since either their absorption from the gut is inadequate and highly variable or else they undergo metabolic degradation in their passage through the gut wall or the liver prior to entering the general circulation. Substances that are either readily oxidized or else hydrolysed in gastric acid may have to be formulated in capsules or enteric coated tablets, whilst some drugs cannot be presented as syrups or elixirs either because of their susceptibility to hydrolysis or their unacceptable taste. The physical properties of others render them unfit for manufacturing processes such as compression into tablets. Sometimes a formulation that was easily prepared in the laboratory or the pilot plant cannot be manufactured economically on a large scale. Unfortunately, this may not be apparent until a late stage in the project when large batches of the active ingredient become available for the first time. If the formulation is then altered, it will be necessary to perform new tests to satisfy the regulatory authorities that safety and efficacy have not been compromised. The delay whilst this is carried out can be considerable.

CAPSULES

The simplest solid oral dosage form to manufacture, manually or on an industrial scale, is the hard gelatin capsule. This merely requires the lower portion to be filled with the pulverized drug and an inert diluent, then capped with the other portion. Volatile or foul-tasting water-immiscible liquids can be presented in soft gelatin capsules. These are prepared in two halves which are sealed by special machines. The advantage of avoiding the problems of unpleasant taste or acid instability of ingredients would seem to make capsules the ideal oral dosage form. Indeed, this is why they are normally chosen for early clinical studies on new drugs. However, capsules cannot be produced as quickly or as cheaply as tablets. This becomes an important consideration when large-scale production is involved.

Lactose or mannitol are suitable diluents for capsules, but other materials can also be selected. Although the diluents are themselves pharmacologically inert, they can have a profound influence on therapeutic action by modifying the rate of release of the active ingredient. This became evident in 1968 when an Australian manufacturer replaced calcium sulphate dihydrate with lactose as a diluent in capsules of phenytoin. Because the original diluent had somehow retarded the rate of release of phenytoin, its replacement caused at least 51 patients to experience toxic effects from enhanced absorption due to faster release of phenytoin in the gut. As phenytoin happens to be a particularly toxic drug, the change in bioavailability became apparent. Its frequently encountered side-effects include gastro-intestinal disturbances such as nausea, vomiting and constipation, as well as central effects such as ataxia, slurred speech, headache, hallucinations, dizziness, insomnia, mental confusion and blurred vision.

COMPRESSED TABLETS

Tablets are made by compressing free-flowing powder fed from a hopper into die cavities in a rapidly rotating press. As the cavities fill, the powder is pressed between two steel punches. The size, shape and markings of the finished tablets are determined by those of the die cavities and punches. This simple description belies the true complexity of the process, which is today based on a rigorous application of the principles of powder technology. Unless the physical characteristics of the powder are appropriate, compression will fail to produce acceptable tablets. It is rare for the raw ingredient to have the necessary physical characteristics, whether it happens to be crystalline or amorphous. This means that biologically inert excipients are incorporated into the powder mass to ensure cohesiveness and other desirable characteristics. As a consequence of this and other possibilities that will now be outlined, as many as twenty or so excipients can be present in a tablet. This means that exceptional care must be taken to ensure the thoroughness and total reliability of the powder-mixing process so as to ensure uniformity of tablet content. Only high-quality, well-maintained equipment can be relied upon to achieve this.

If a crystalline drug were compressed on its own, it would probably crumble either on removal from the machine or when subsequently handled. In order to avoid this, an adhesive binding agent is incorporated into the formulation. This will consist of a mucilaginous solution prepared from starch, gelatin or an appropriate sugar. A variety of naturally occurring gums may also be employed. When it is necessary to delay the disintegration of the tablet on contact with body fluids, as with throat lozenges or antacids that are to be sucked, acacia is employed instead as it is a stronger adhesive and does not dissolve so readily. Frequently, a lubricant such as talc, stearic acid or its magnesium salt has to be added to avoid adherence of the powder to the tabletting machinery.

In cases where a drug is particularly potent and the tablet is thus likely to be inconveniently small, an inert diluent has to be incorporated to afford sufficient bulk for easy handling by the patient. Calcium salts (sulphate, phosphate), sodium chloride, sucrose or lactose, and starch are typical fillers. Problems also arise with drugs of low potency, for then the excipients constitute a relatively small proportion of the finished tablet. In such cases, the physical properties of the active ingredient may not be ideal for production of tablets by direct compression. It then becomes essential to subject the tablet powder to a special granulation process. This is normally done by moist granulation, which involves making the powder into a paste by mixing with a viscous solution of an appropriate binding agent. After rubbing through a sieve, the paste is dried to produce granules ready for compression into finished tablets. An alternative process is required for preparing granules of water-sensitive drugs, namely that of dry granulation. This involves making crude tablets by direct compression, followed by crushing and grinding to obtain satisfactory granules.

Any drug with very low water-solubility presents particular problems for the formulation pharmacist. Both its particle size distribution and shape, as well as the degree of compression of the granules containing it and the excipients, can profoundly influence the rate at which dissolution in body fluids takes place. Faster dissolution can often be achieved by incorporating disintegrants which swell on contact with water. These are natural products such as starch, gelatin, agar, alginates or modified cellulose derivatives, e.g. carboxymethylcellulose. Some formulations also include surfactants as wetting agents to enhance dissolution of those drugs which tend not to be wetted by water. Generically equivalent preparations containing the same amount of active ingredient, but with either no, or different, disintegrants or surfactants, may be available from an alternative manufacturer. As a result, the bioavailability of the active ingredient may be different. This does not seem to be a serious problem with those drugs that are readily soluble in water, unless they are formulated as slow-release preparations.

Excipients may have to be incorporated into a tablet formulation to enhance stability. These include antioxidants and even antimicrobial preservatives for tablets with a high moisture content. Flavouring and colouring agents are also often required, the permissible range of which is strictly regulated.

Figure 5. Tablet manufacture. *Reproduced by permission of the Glaxo Group of Companies.*

An elegant, though costly, formulation is the effervescent tablet containing citric acid and sodium bicarbonate, which do not interact to liberate carbon dioxide so long as the tablet remains dry. The widespread use of foil wrapping has facilitated the production of this type of tablet by ensuring that storage under dry conditions is easily achieved.

Tablets are often coated in order to improve appearance, facilitate handling, mask an unpleasant taste, afford stability against gastric acid, or prolong the release of the active ingredient. Traditionally, this has been done by first sealing tablets with an alcoholic solution of either shellac, a cellulose derivative, or polyvinyl acetate phthalate. This stops penetration by the aqueous sugar solution subsequently added to build up layers of coating. Before the sugar coatings can be built up, a special gummy subcoating has to be applied to the sealed tablet. Each sugar layer is then slowly hardened by a current of hot air whilst rotating in a stainless steel pan. This tedious process, which creates a bottle-neck in the production line, calls for considerable skill and is expensive. It is increasingly being replaced by film coating techniques. These gained popularity because they are much faster as polymer coating is dissolved not in water, but in volatile solvents. This may either be sprayed on tablets suspended in an air current, or else applied in much the same way as a sugar coating. The hazard caused by the inflammability of some volatile solvents can now be avoided by the use of aqueous systems in film coating, thanks to the emergence of highly efficient drying procedures.

Cellulose-based coatings can have varying degrees of solubility in gastro-intestinal fluids, whilst being highly resistant to moisture and oxidation. Unlike sugar-coated tablets, there is little difficulty in preparing film-coated ones for storage in humid conditions.

LIQUID ORAL DOSAGE FORMULATIONS

Mixtures consist of either water-soluble drugs in simple aqueous solution or insoluble compounds suspended in water to which an appropriate suspending (thickening) agent has been added to ensure an even dispersion of the drug when the bottle is shaken immediately before administration. Failure to achieve this would mean that the wrong dose is administered. Syrups, prepared by dissolving drugs in 66 per cent sucrose solution, are normally intended for mixing with other ingredients, as in linctuses that are sipped slowly. Problems are created by some moderately water-soluble drugs that tend to crystallize out of solution on storage. To prevent this, these may be dissolved in either alcohol, glycerol or propylene glycol before a flavoured syrup is added to form an elixir. This type of formulation can also be used to mask the unpleasant taste of some substances. When an oil has to be formulated for oral administration its unpalatability may be overcome by emulsifying it to form a fine dispersion in water, stabilized by the addition of a surfactant or traditional emulsifying agent. A basic requirement for all of these formulations is that the drug is stable under the conditions of storage likely to be encountered.

Cellulose derivatives, which are often used as thickening agents in liquid formulations, affect gut motility. Their presence or absence from a formulation can therefore influence the rate of absorption, especially in paediatric formulations. This could result in differences in bioavailability between generically equivalent preparations.

PARENTERAL PRODUCTS

Injections are amongst the most critical of all formulations to manufacture. Since their administration by-passes many of the defensive barriers of the body to foreign substances, considerable harm may result if injections are contaminated by microorganisms. In the past, terminal sterilization of the sealed injection ampoule was considered satisfactory, but the modern approach conforms with the concept of building quality assurance into the production line by avoiding unnecessary contamination from either the air or the water supply, both of which can be sterilized. Sophisticated air filtration and the maintenance of production suites under positive air pressure help to avoid air-borne contamination. A series of air-locks may be utilized to minimize the likelihood that the sterile air is contaminated by the entry of the production

Figure 6. Sterile filling of ceftazidime injection ampoules. *Reproduced by permission of the Glaxo Group of Companies.*

staff. The filling of ampoules is preferably done by machines bathed in ultraviolet light or, if hand filling is adopted, in laminar flow cabinets under currents of sterile air that prevent operator-borne contamination. In some factories, staff are not only clad in sterilized gowns, hoods, masks and gloves, but even wear lead-weighted, covered footwear to slow their movements in order to prevent disturbing the surrounding atmosphere!

Pain and irritation may be caused either directly by the injected drug itself or by its precipitation due to the change in the environment of the injection solution as it makes contact with the body fluids. This is most likely to happen when weakly acidic drugs have to be formulated at a pH between 8 and 11 in order to remain in the water-soluble ionized form. On contacting the body fluids buffered at pH 7.4, such drugs are precipitated as the insoluble non-ionized species. Similar problems may arise with very weak bases formulated at a pH between 3 and 6. As with acids, this is only likely to happen when the undissociated species is insoluble in water.

The stability of injections is an important consideration since even traces of impurities can be harmful. The most frequent problems arise from sensitivity to either hydrolysis or oxidation. This can be illustrated by considering the difficulties encountered when formulating adrenaline injections. Adrenaline can be dissolved in water, but the slightly alkaline solution that results is rapidly oxidized to inactive adrenolutin. The problem cannot be overcome simply by adjusting the pH to neutrality since a different oxidation product, adrenochrome, is then formed. If it is necessary to employ an adrenaline solution with a neutral pH, an antioxidant must be added. Care should be exercised in choosing one, for the widely used sodium metabisuphite can react chemically with adrenaline above pH 4. In neutral eye drops of adrenaline this problem has been overcome in two ways. Firstly, a borate buffer system was chosen since boric acid formed a soluble complex with the oxidizable catechol hydroxyl groups in the adrenaline. This prevented their conversion to a quinone, whilst still being capable of releasing adrenaline in the eye. Secondly, 8-hydroxyquinoline was selected as a non-irritant antimicrobial preservative because it reacted with oxidation-promoting trace metals to form a metal-chelate. These elaborate measures were not necessary if the adrenaline was formulated between pH 2.8 and 3.8. This had been shown to be the optimal pH range to prevent oxidation of adrenaline. When, for example, local anaesthetics are formulated with adrenaline added as a vasoconstrictor to prolong anaesthesia, the pH is reduced to around 3. Any lower pH would be irritant, so the fact that adrenaline undergoes racemization to the less active dextrorotatory isomer when formulated below pH 2 is only of academic interest. The sensitivity of many other bases to oxidation can also be diminished by formulating them at an appropriate pH. The same applies to some substances that undergo hydrolysis. In this connection, consider the formulation of physostigmine eye drops. These were easily recognized as being unstable because they soon turned pink. The colour did not form when bisulphite was added as an antioxidant. This was at first misunderstood, but later it was realized that the prime problem was the hydrolytic cleavage of the carbamate ester function to form inert eserinol. This decomposition product turned red on oxidation, a secondary process that was prevented by the antioxidant. It was then discovered that the bisulphite also lowered the pH to below 6, and thus retarded the initial hydrolysis. The same effect was achieved simply by adjusting the formulation pH without the need for incorporation of bisulphite.

CONTROLLED RELEASE FORMULATIONS

Control of the release of the active ingredient from dosage forms has become a major focus of attention in the pharmaceutical industry, with novel active delivery systems now being carefully scrutinized. These include tablet-sized devices which pump into the gut a regulated amount of drug through a fine, laser-drilled aperture in a polymeric membrane for a predetermined period of time. The driving force for this process is the osmosis generated by diffusion of fluid into the device. Unfortunately, an early example of this type of active delivery system containing the antirheumatic agent indomethacin, *viz*. 'Osmosin', had to be withdrawn from the market in 1983 because of an unacceptable incidence of damage to the gut wall.

Ideally, the rate of release of the medicinal ingredient from a capsule or tablet should be slower than the rate at which it is absorbed from the gut into the circulation. This will mean that the drug is absorbed soon after it is released, entry of it into the circulation being determined more by the nature of the formulation than by the vagaries of the human digestive system. In practice, this is rarely achieved. Furthermore, the period during which the tablet or capsule resides within the region of the gut where drug absorption occurs fluctuates widely not only between different people, but also for any one individual. Whilst this may average something in the order of $2\frac{1}{2}$ hours, it can sometimes differ from this by 1 hour, shorter or longer. This, in turn, is reflected in the amount of drug that will be detected in the plasma. Obviously, diet exerts an effect, but so do other physical and emotional factors which influence the degree of peristaltic propulsion of the contents through the small intestine. The situation becomes still more complicated through the variability in the gastric emptying rate according to time of day. Furthermore, there seems to be a diurnal variation in the activity of metabolic enzymes, whilst renal tubular reabsorption of ionizable drugs is influenced by the variation in urinary pH with both time of day and diet (it can be as low as 5.6 before breakfast and after the consumption of acidic beverages such as fruit juices and cola drinks). Little wonder that once-daily dosing is considered preferable to the traditional four times a day. Unfortunately, this is often unattainable with rapidly metabolized drugs despite the use of sophisticated formulation techniques or pro-drugs which are slowly metabolized to release the active moiety into the circulation.

Sustained release of the ingredients of capsules is relatively easy to achieve by incorporating not a simple powder, but a mix of granules with varying dissolution characteristics. This is done by coating the granules in much the same way as tablets are coated. To achieve a slow release in tablets, they have to be made with either a strong binding agent or a special coating, otherwise the drug must be embedded in a permeable wax matrix from which it will slowly be leached out. For these processes to be reliable and depend solely on the nature of the formulation and not on haphazard absorption from the gut, the drug should be highly water-soluble so that it will be absorbed soon after release. This is not always feasible, although reduction of particle size will sometimes enhance dissolution. With coated tablets, delayed rather than slow release occurs since the tablet behaves like any other once the coating has dissolved. There are problems, too, when a drug is embedded in a wax matrix. As time passes, the rate of release inevitably slows since any drug remaining in the insoluble matrix is to be found in the least accessible portion. This adds a further complication to the picture of drug absorption. Attempts to overcome the problem often fail to ensure a constant release of drug. The best way to guarantee any level of predictability with regard to release rates is to ensure strict control over the quality and blending of the many excipients that can be incorporated into the formulation.

When ethylhydrocupreine was used between the two world wars as an oral antibacterial in the treatment of pneumonia it had to be formulated as the insoluble free base in order to slow down the rate of absorption from the gut and thus avoid attainment of toxic blood levels. This use of slow release formulation remains rare. Instead, such formulation is mainly employed with the intention of sustaining the action of a drug so as to minimize the frequency of dosing. However, the brevity of the transit time of drugs through the absorptive region of the small intestine casts doubts on the wisdom of relying on slow release formulations to prolong the action of drugs with short half-lives. It has also been asked why it is necessary to have these formulations for drugs which persist in the circulation for longer periods, e.g. psychopharmacological agents and antihistamines.

In view of all the uncertainties, it is hard to understand why so many controlled release formulations are available. Have marketing considerations been responsible for this state of affairs? It has been hinted darkly that since the release rate profiles of different brands cannot be precisely matched in bioavailability tests, slow release formulations have been marketed to deter generic substitution by rival products. However, there has been controversy over whether the variations between different brands of the same drug are of greater significance than the variability in drug absorption profiles between different patients or even in any one individual swallowing a sustained release formulation on different occasions. It is probable that differences between brands are only important in drugs with a narrow therapeutic window. There are, for example, several brands of slow release formulations of theophylline on the market. Their release rates vary widely, a feature that must give cause for concern as the margin between minimum effective blood level and the toxic level of this drug is small. In the United States, the Food and Drug Administration has adopted an ultra-cautious approach by requesting manufacturers to submit data on the dissolution rates of all solid dosage formulations.

It is not just in the digestive tract that slow release formulations are used. Injections may be formulated as fine suspensions of insoluble drug for intramuscular or subcutaneous administration, as is the case with insulin zinc suspension and several vaccines. Considerable prolongation of drug action can sometimes be attained by the use of specially formulated depot injections which deposit fine globules of an oily drug solution around the site of its injection into a muscle. The technique is obviously limited to highly potent agents. As only very lipophilic substances will be retained in the slowly dispersing oil droplets for adequate periods of time, this procedure has been restricted in its application to drugs such as the steroid contraceptive medroxy-progesterone acetate and ester derivatives of tricyclic tranquillizers that are administered once or twice monthly as maintenance injections in schizophrenia, e.g. fluphenazine and flupenthixol.

The peaks and troughs in tissue levels of pilocarpine following its intermittent administration in eye drops for the treatment of glaucoma can now be avoided by placing in the eye a small polymeric membranous device known as an 'Ocusert'. It is claimed by the manufacturer that a sustained release of controlled amounts of the drug over seven days avoids the annoying changes in refraction that patients can experience for the first hour or two after administration of conventional eye drops several times a day.

When salbutamol was introduced for use in asthma its resistance to enzymatic destruction by catechol O-methyltransferase provided scope for it to be utilized not only therapeutically but also prophylactically since it persisted in bronchial tissue after inhalation. Unlike isoprenaline, which it superseded, the free base form of salbutamol was not extremely sensitive to oxidation. This meant it could be inhaled from a metered aerosol in this form which was much less water-soluble than any salbutamol salt. The resultant slow dissolution further prolonged the action on the lung by creating an undissolved reservoir of the drug, permitting 4-hourly dosing for prevention of bronchoconstriction. Nor is this the only example of formulation techniques being used to obtain more effective delivery of a drug to the lungs. The success of sodium cromoglycate has depended on a special device which permits micronized particles of this very insoluble drug to be inhaled in the form of an insufflation.

OINTMENTS AND CREAMS

Dermatological preparations constitute perhaps the most unreliable of all pharmaceutical products since it is rather difficult to control either the amount applied to the skin or the rate of release of the active ingredients. Penetration of drugs through the skin is enhanced by hydration of the stratum corneum. This can be achieved by skin occlusion to prevent evaporation of surface water, or to a much lesser extent by a water-repellant ointment base. Polyethylene glycol gels can also assist penetration markedly.

Not only penetration, but also release of the active medicament from the base has to be considered. When rapid release of the medicament for deep penetration is desired, an emulsified cream base is selected. If a slow release rate is wanted, a greasy ointment base is to be preferred. However, this presupposes that the chemical properties of the drug permit such a simple choice. Salts of some acids or bases may interact adversely with emulsion bases and thus fail to penetrate the skin. Indeed, any alteration or dilution of a proprietary cream formulation can destroy its bioavailability. Unless experiments on humans have already shown that absorption is not affected, prescribers should be wary of altering such formulations.

TRANSDERMAL MEDICATION

The ability of some highly potent drugs to pass through the skin and enter the circulation has been recognized for a long time. Belladonna preparations were occasionally able to relieve spasm when applied to the surface of the body in an appropriate area, whilst glyceryl trinitrate ointment relieved anginal attacks.

The latter provided a reservoir for sustained release of the otherwise short-acting drug. The problem with this approach was that, as with all ointments, control over the amount applied was hopelessly inadequate, leading to an unacceptable variation in clinical response. A modern development of this approach is a sophisticated transdermal patch which ensures a constant release of the drug into the circulation over a period of time which can extend to seven days. An occlusive backing is employed to achieve high penetration of the drug which is precisely released from a special polymeric material. Amongst the first agents to be considered for administration in this way are hyoscine (a constituent of belladonna) for motion sickness, and glyceryl trinitrate. Several manufacturers are currently in the process of introducing their own brands of transdermal glyceryl trinitrate delivery systems, but unfortunately these all differ in their release rates and thus pose a potential hazard to the patient who switches from one brand to another.

III

Quality Assurance

Concern about the quality of medicines dates back at least to the Ancient Greeks. In his *Historia Plantarum*, the fourth century BC writer Theophrastus was concerned about the source, manner of collection, and storage of medicinal plants. The problem of adulteration was discussed four centuries later by Dioscorides, who cited 40 examples including the adulteration of opium with plant juices. He explained how adulterated drugs could be detected by examining either their taste, smell, physical properties such as solubility in water, or their behaviour when placed in a flame. During the medieval period, the premises of Arabian apothecaries were regularly inspected by government-appointed officials who closely watched over the compounding of medicines and checked the accuracy of the weights and measures. This practice was followed in Germany by Frederick II, who arranged for the routine inspection of the medicines compounded by apothecaries. If a medicinal preparation proved lethal, the apothecary was liable to face the death penalty! Controls over the activities of apothecaries followed in Italy and France, and in England,

Figure 7. The inspection of a pharmacist's shop in the seventeenth century. Engraving from J. Michaelis, *Opera Medico-chirurgica. Reproduced by permission of the Wellcome Institute Library, London.*

where the City of London appointed its first drug inspector in 1423. Henry VIII introduced an Act in 1540 which enabled the London College of Physicians to appoint its own inspectors.

The appearance of formularies and pharmacopoeias in Western Europe soon after the introduction of the printing press helped to improve the quality of medicines by establishing standards, albeit ill-defined, against which products could be compared. Proper control over the quality of medicines remained unattainable until accurate analytical methods could be developed to detect and quantify the amount of active ingredients and reveal, too, the presence of impurities. This was not achieved until well into the nineteenth century. In the United States, the first statute to control the quality of medicines was introduced in 1848, following an incident involving adulteration of quinine destined for the army. After the Sale of Food and Drugs Act of 1875 had come into operation in the United Kingdom, analytical techniques developed sufficiently to ensure that deliberate adulteration of drugs became a thing of the past. The problem that remained was that of unintentional adulteration due either to negligence or to the inadequacy of manufacturing processes.

The term 'analysis' was introduced by Robert Boyle, the Oxford scientist who conducted the earliest systematic investigations in the field of qualitative analytical chemistry. He made accurate measurements of the physical properties of different substances, and in 1690 published his *Medicina Hydrostatica: or Hydrostaticks Applied to the Materia Medica*. It extolled the virtues of measuring specific gravity, but this was of no value in the assessment of the herbal products which were then dominant. Boyle also developed the first chemical, as opposed to physical, methods of analysis. For instance, he employed vegetable juices as indicators to detect acids and alkalis, and demonstrated that a variety of reagents could precipitate certain chemicals from their solutions.

In the late eighteenth century, French scientists made notable advances in analytical chemistry as applied to medicine. Antoine Fourcroy improved upon the standard procedure of assaying the medicinal value of mineral waters, which had involved evaporating them to dryness and weighing the residue. Instead, he introduced specific chemical reagents that indicated which minerals were present. In 1791, he published an analysis of St Lucia and St Domingo barks, which had been recommended as substitutes for cinchona bark in the treatment of malaria. This was for many years considered to be a model of vegetable drug analysis and it stimulated others to examine cinchona and opium, the two most important drugs then in use. After the French Revolution, the responsibility for analysing medicines was vested in the Société de Pharmacie and the École Supérieure de Pharmacie, the latter opening in Paris in 1803. Its first director was Nicolas Vauquelin, a close associate of Fourcroy. He was an outstanding analytical chemist and did much to establish gravimetric methods for the quantitative assay of minerals. Vauquelin responded to growing criticism in medical circles over the variable quality of plant products by encouraging his faculty members and their students to devise new methods of plant analysis. These involved extracting the unidentified pharmacologically active principles so that they could be quantitatively determined by weighing. This led to the isolation of several important alkaloids, followed by methods for their accurate determination. It also resulted in the introduction by Magendie of pure alkaloids as medicinal agents—the greatest single step ever taken towards improving the quality of drugs. Instead of relying on herbal products of highly variable quality, physicians obtained purified alkaloids from manufacturers who vied with each other in proclaiming the high standards of their products.

A major stimulus to the development of quantitative methods of analysis was the atomic theory propounded by John Dalton. This pointed the way for Berzelius to elaborate his concept of equivalent weights, which enabled analysts to calculate the amount of one substance by reacting it with a measured quantity of another. In 1828, Gay Lussac, the doyen of the Parisian chemists, then perfected Vauquelin's technique of volumetric analysis, introducing alkalimetry to determine the quantity of acid in a solution. Although the initial response to this development was slow, later in the century volumetric analysis came into universal use for the assay of drugs. It remained unrivalled until after the Second World War when the rapid development of electronic instruments gradually brought quantitative spectroscopic techniques to the fore. These had their origin in work done in the 1870s when photographic methods had first shown that solutions of many organic compounds absorbed light and that the wavelength of maximum absorption was specific for each compound. In 1885, W.N. Hartley investigated the absorption of ultraviolet light by alkaloids, and he came to the conclusion that their spectra were sufficiently characteristic to offer a simple means of ascertaining both identity and purity.

The popularity of modern methods of instrumental analysis is due to the fact that they often do not require a tedious separation of the material being assayed. Nevertheless, separation techniques have themselves advanced astonishingly through the use of a variety of chromatographic methods of exquisite sensitivity. Spectroscopic and chromatographic instruments can now easily distinguish between a drug and its impurities with chemical structures so similar that they cannot be distinguished by the reagents traditionally employed in volumetric analysis. Indeed, using methods such as combined mass spectrometry–gas chromatography, impurities can routinely be detected at levels far below those at which the safety of drugs would be affected. These modern developments ensure that regulatory authorities can be satisfied that pharmaceutical products really do contain the correct amount of the stated active compound within precisely definable limits. The stability of the product can also be reliably established since traces of decomposition products are readily detectable throughout the period of the permitted shelf-life.

The limits of modern analytical techniques are reached when it comes to the examination of the complex mixture of ingredients present in many finished pharmaceutical products, particularly tablets, capsules, creams and ointments.

Recognizing this, guidelines were drafted in 1967 by the US Pharmaceutical Manufacturers' Association in order to ensure the highest possible standards of quality throughout the industry. These formed the basis of regulations subsequently issued by the Food and Drug Administration, *viz. Current Good Manufacturing Practice in Manufacture, Processing, Packing or Holding of Human and Veterinary Drugs*. These provide quality assurance for the entire process of drug manufacture and distribution, from procurement of raw materials right through to handling by wholesalers and pharmacists. All of this was over and above reliance on retrospective quality control by means of routine sampling and analysis. The guidelines for good manufacturing practice published by regulatory authorities on both sides of the Atlantic and within Europe are broadly similar so as to satisfy the demands of the international trade in pharmaceuticals. They have helped to advance the slow but steady progress towards international harmonization of pharmaceutical licensing now taking place, as exemplified by the Pharmaceutical Inspection Convention facilitating mutual recognition of factory inspections conducted in participating European countries.

Quality assurance is an integral part of modern pharmaceutical production, requiring much more than merely checking the quality of the individual ingredients that are to be compounded into finished products. All stages of production are regulated, monitored and validated with a view not only to avoidance of contamination but also to minimizing the risk of human error. For instance, the movement of personnel within a factory may be controlled to ensure there is no chance of materials from one production line being negligently transferred to another. Nevertheless, mistakes can occur. In the summer of 1985, several asthmatic patients suffered severe respiratory distress necessitating artificial respiration after they had received injections from ampoules labelled as containing hydrocortisone sodium succinate. Somehow, the wrong labels had been attached to vials of vecuronium, a neuromuscular blocking agent produced by the same manufacturer.

Past mishaps have demonstrated that stringent controls over the production of sterile products are necessary. In 1971, inadequate autoclaving in an English factory producing parenteral infusion fluids was only detected after the condition of several patients markedly deteriorated. The government instituted an inquiry which found that one third of the batch concerned was contaminated with bacteria. This led to heightened demands for stricter checks on autoclaving machines. Since then, there has been a general reassessment of what constitutes good manufacturing practice for sterile products. All processes from the initial cleansing of apparatus through to the filling of ampoules and their terminal sterilization are now conducted in a series of rooms with increasing levels of control over the entry of airborne microorganisms. Extremely high standards of cleanliness are a *sine qua non*. In Sweden, matters are taken further by an insistence that all ingredients must be subjected to an initial microbial count that ensures their suitability for use in the preparation of injections. It has become generally accepted that, whenever possible, terminal batch sterilization by heat treatment must be preferred to the removal of microorganisms by aseptic filtration. The efficiency of the latter process can only be checked by sterility tests, whereas that of heating the finished product can, because it is a batch process, be monitored in a variety of ways, including the use of microbial indicators.

The concept of quality assurance extending to cover even the manner in which products will be handled by patients is nicely exemplified by the current awareness of problems surrounding the need for sterility in both eye and ear drops. In the past, the most that could be expected was that multidose vials of these would contain a preservative and be supplied to the patient in a sterile condition. Often, pharmacists were (and occasionally still are) called upon to prepare these extemporaneously, relying on the incorporation of a bacterio-static agent and boiling at 100°C when an autoclave was unavailable. Concern over microbial contamination after the opening of the container by the patient is now persuading manufacturers and regulatory authorities that sterile unit-dose products are superior.

THE STABILITY OF PHARMACEUTICALS

Any pharmaceutical product may undergo chemical or microbiological de-terioration after it has been dispatched from the factory. Manufacturers conduct a variety of tests to ascertain how long their product remains suitable for therapeutic application. These tests are initiated as soon as practicable once the favoured formulation of a new drug has been devised. They are carried out not only on the active ingredient but also on the finished product. Accelerated storage tests at extremes of temperature and humidity are of much value, for they permit early modification of the formulation before clinical trials have proceeded too far. In any case, regulatory authorities will not permit a trial to proceed unless the formulation used has been shown to be stable throughout the trial period.

It is possible by careful consideration of packaging requirements to avoid the need for formulation changes. For instance, the sensitivity of an injection solution to atmospheric oxidation may be heightened by light, alkalinity and the presence of certain trace metals. Rather than incorporating an antioxidant in the formulation, it may be sufficient to present the injection in ampoules wrapped in light-proof paper and sealed under nitrogen, the glass from which the ampoules are made being free of the offending trace metals and excess alkalinity.

Many liquid formulations contain antimicrobial preservatives. However, microbial contamination can affect the stability of any pharmaceutical product, especially those compounded with carbohydrate excipients such as starch, acacia or tragacanth. Solid dosage forms may be exposed to conditions of high humidity which turns these excipients into excellent microbial growth media! The attraction of syrup as a vehicle for preparing liquid concentrates for use in the dispensary goes beyond its palatability; it will not sustain the growth of bacteria unless diluted with water—hence the reluctance of manufacturers to abandon syrup-based formulations despite the damage to the teeth of children and the hazard to diabetic patients.

IV

Drug Toxicity

The nature of the problem of drug toxicity has been changed by the introduction of synthetic drugs. This is because the process of screening novel organic compounds so as to detect those with enhanced selectivity of action has ensured that most of the favoured candidate drugs will have relatively low levels of toxicity. Often, serious problems only arise after prolonged administration of a drug. This contrasts strikingly with the situation in the past. If the role in nature of the plant poisons which then constituted the mainstay of therapy really was to repel foraging predators, it is understandable why the onset of their toxicity was rapid. Whether or not that interpretation of their role is valid, it can hardly be disputed that at the dose levels traditionally employed in the past, drugs such as quinine, digitalis and colchicine poisoned many patients.

The clinical and commercial success of a new drug is as likely to be determined by its toxicity as by its efficacy. As the safety of new drugs improves, the general public becomes conditioned by the media to expect something approaching absolute safety in the medicines they consume in ever-increasing quantities. Such safety is unattainable because the similarity of the biochemistry of all cells in the body ensures that if enough drug is administered, the function of organs other than the target one will also be altered. Nevertheless, still higher standards of safety can be achieved, but the price to be paid for this in terms of preclinical development and safety testing needs to be balanced against other expenditure on health care. Whilst serious adverse drug reactions are presently seen in only a very small proportion of patients, minor ones are quite common. As part of the Boston Collaborative Drug Surveillance Program, it was established that nearly one-third of patients experienced at least one adverse drug reaction during their stay in hospital. As this study dealt with people whose illnesses were severe enough to require hospitalization, it must be seen in perspective. Thus, the Boston investigators found that the drugs causing multiple deaths most frequently were anticancer agents, digoxin and heparin. Amongst the general population, the commonest adverse effects of drug therapy are drowsiness, nausea, dizziness and dry mouth. This is probably a reflection of the widespread use of psychopharmacological agents. The single drug causing the highest incidence of adverse reactions in recent years seems to have been ampicillin, with most reports citing the allergic uticaria it produces. The Boston study gives an indication of the incidence of serious adverse drug reactions in the general population by its finding that 3 per cent of hospital admissions were due to adverse drug reactions.

Thorough toxicity testing and pre-marketing clinical trials ensure that any compound causing serious toxicity in, say, more than one in a thousand patients simply will not be marketed unless it is a life-saving agent which is absolutely essential because no better alternative exists. To achieve this state of affairs, it now takes ten years or more to bring a promising new drug to the market.

Adverse drug reactions can be caused by direct damage to the tissues by exposure to excessive amounts of the drug or indirectly by allergic phenomena. A distinction should be made between toxicity and allergic hypersensitivity because the former may be remedied by dosage adjustment, whereas the latter requires withdrawal of the offending agent. It is also useful to distinguish between adverse reactions and side-effects, the latter being an inevitable consequence of the pharmacological action when normal doses are administered. A common example would be the annoying anticholinergic side-effects caused by the lack of selectivity of phenothiazine tranquillizers and many other drugs, or the postural hypotension induced by many antihypertensive agents. Secondary effects are those which arise from functional changes induced by drug therapy, such as the consequences of alteration of the gut flora through the oral administration of antibiotics.

IDIOSYNCRASY AND INTOLERANCE TO DRUGS

'Idiosyncrasy' is the term used to describe a qualitatively unusual toxic response to a drug, this not being due to allergy. The best-known example is the fatal haemolytic anaemia observed in several patients of African and Mediterranean extraction during the initial clinical trials of pamaquin, the first synthetic antimalarial drug. After the Second World War, pamaquin was replaced by the much safer primaquine. However, cases of fatal haemolytic anaemia still continued to appear. Ultimately, this was associated with low activity of glucose-6-phosphate dehydrogenase. This enzyme protects cells against oxidative stress caused by exposure to oxidants such as synthetic antimalarials, synthetic antibacterials (viz. chloramphenicol, dapsone, nalidixic acid, nitrofurans, 4-aminosalicylic acid and sulphonamides), aspirin, phenazone, amidopyrine and probenecid. It has been estimated that around the world there are one hundred million people with glucose-6-phosphate dehydrogenase deficiency.

Intolerance is quite distinct from idiosyncrasy insofar as it is a quantitatively rather than a qualitatively unusual response to drug therapy. It is seen when patients react badly even to modest doses of certain drugs. Such intolerance may have a genetic basis which manifests itself as an abnormally low activity of a specific drug-metabolizing enzyme, as with suxamethonium (succinylcholine). This short-acting neuromuscular blocking agent was introduced in 1949. Its brief duration of action was due to rapid enzymatic hydrolysis by plasma pseudocholinesterase. Unfortunately, in a small proportion of patients there was an abnormally low level of pseudocholinesterase activity due to their enzyme being atypical in its structure. In such genetically predisposed individuals, the use of suxamethonium resulted in prolonged apnoea from paralysis of the respiratory musculature. They sometimes had to be artificially ventilated for many hours. Of less serious consequence was the finding that about 5 per cent of the British population suffered side-effects from the hypotensive agent debrisoquine because of inefficient conversion of the drug to inactive 4-hydroxydebrisoquine. This was first noticed when it was found that such patients could be maintained on doses of less than 10 mg twice daily, whereas others could require more than six times as much.

Aspirin is a hazardous substance for the 10–20 per cent of asthmatics who are sensitive towards it. In these patients, severe bronchospasm or even fatal status asthmaticus can follow a single dose. Several chemically unrelated non-steroidal anti-inflammatory agents such as indomethacin, ibuprofen and fenoprofen also produce similar effects. This, together with the high incidence of the intolerance amongst asthmatics, and the failure to detect antibodies, has raised doubts as to whether a straightforward allergic response is involved, even though urticaria and other allergic reactions to aspirin are not infrequent. All the drugs affecting asthmatics in this manner interfere with arachidonic acid metabolism by blocking the cyclo-oxygenase pathway. It seems possible that this favours increased production of the slow-reacting substance of anaphylaxis (SRS–A) through the alternative lipoxygenase pathway. This might account for the intolerance in asthmatics as SRS–A is known to be a mediator of allergic reactions.

CUMULATIVE DRUG TOXICITY

Toxicity can be caused when traces of a previous dose of a drug have not been cleared from either the general circulation or from a specific organ. When only very minute amounts of material accumulate, months or years may pass before toxic effects become apparent. Examples of this will be cited in the following pages. However, a much shorter period can be involved, as was the case when the United States suffered a major drug disaster which compounded that inflicted by the Japanese at Pearl Harbor. Casualties were anaesthetized with the recently developed intravenous agents hexobarbitone and thiopentone. These agents were short-acting because a substantial proportion of the drug was quickly metabolized, as a consequence of which brain levels of the drug fell below the anaesthetic threshold. The cumulative effects of the still unmeta-bolized drug led to a lethal plasma level when repeated injections were administered. This had not been fully appreciated before Pearl Harbor, with the consequence that many lives were unnecessarily lost during the turmoil that ensued.

In 1937, more than one hundred Americans died after being treated with Elixir of Sulphanilamide, which incorporated 75 per cent aqueous diethylene glycol as a solvent for the active ingredient. Tests had previously confirmed that the acute toxicity of diethylene glycol was lower than that of glycerol, a similar vehicle used in many formulations. Unfortunately, it had not been realized that repeated administration of diethylene glycol caused progressive renal damage. When the scale of the disaster was realized, Congress introduced the Food, Drug and Cosmetics Act to control the testing and marketing of new drugs. Important lessons learned from this episode were that chronic toxicity testing was as necessary as acute toxicity testing, and new formulations of existing drugs needed to be scrutinized just as carefully as novel compounds.

Soon after isoniazid was introduced in 1952 as an antituberculous drug, it was found that some patients experienced adverse effects because of high blood levels. This was attributed to slow metabolic detoxification by acetylation, resulting in accumulation of the drug. The biological half-life of isoniazid was shown to be 45–80 minutes in fast acetylators, and 140–200 minutes in slow acetylators. Since then, it has been realized that the facility with which the body acetylates certain nitrogenous compounds is genetically determined, there being two phenotypes. Marked differences exist between racial groups. Fast acetylators are predominant amongst the Japanese and the Eskimos, whereas slow acetylators are common amongst Egyptians. Just over half of the British population exhibit the recessive trait of slow acetylation. Slow acetylators are not only more likely to experience cumulative toxicity from isoniazid if the dose is not appropriately adjusted, but are also exposed to a slightly greater risk of drowsiness and blurred vision from toxic levels of phenelzine, blood disorders after either sulphasalazine (salicylazo-sulphapyridine) or dapsone, or a condition resembling systemic lupus erythematosus caused by hydralazine. Fast acetylators are more likely to encounter hepatoxicity from isoniazid, or crystalluria due to deposition within the kidney tubules of insoluble acetyl derivatives of certain older sulphonamides such as sulphadiazine.

Following the recent benoxaprofen incident, there has once again been a heightened awareness of the problem presented by the accumulation of some drugs, especially in elderly patients. Benoxaprofen was first launched in the United Kingdom in March 1980, being strongly promoted as an antiarthritic agent. Aside from the contentious issue of whether its clinical action had differed significantly from that of other agents, benoxaprofen did offer the opportunity for once-daily dosing because of its resistance to metabolic degradation. This was, however, to prove its undoing. The 15th International Congress of Rheumatology, held in Paris in June 1981, was told that benoxaprofen was slowly excreted, its biological half-life being as long as 4 days in elderly patients. Previously, the half-life was considered to have been 33 hours in man, justifying the convenience of once-daily dosage. The implications of this prolonged half-life in elderly patients were not obvious then, but with the benefit of hindsight they most certainly are now. The issue centres on the recognition of toxicity, for only if a drug is relatively toxic does cumulation as a consequence of poor renal function constitute a problem. Examples of this include streptomycin and other ototoxic aminoglycoside antibiotics, as well as digitalis. Although reports of photosensitivity and nail damage had been received, benoxaprofen was not at that time thought to be any more toxic than other non-steroidal anti-inflammatory agents. Its manufacturer mentioned the need for dosage reduction in a pamphlet issued three months after the Paris symposium, but the importance of this seems to have been generally overlooked. The outcome might have been different if more information about the effects of benoxaprofen in old people had been available. Apparently, 52 patients over the age of 65 received the drug in the early clinical trials. This was by no means unusual since, until then, such trials were not expected to have been carried out on elderly patients. When, in February 1982, Dr Hugh Taggart of Queen's University in Belfast reported the deaths of five elderly patients who had received benoxaprofen, urgent inquiries ensued. The Committee on Safety of Medicines withdrew its Product Licence in August of that year, by which time 83 fatalities amongst three-quarters of a million patients who took the drug had occurred. Most died from renal or hepatic failure.

Troublesome side-effects can be caused in some patients by slowly metabolized drugs in which there can be a wide variation in the half-life. These include amylobarbitone (amobarbital), chlorpropamide, diazepam, digoxin, haloperidol, lithium, methaqualone, phenytoin and warfarin.

The mere fact that the deadliest chemical warfare agents are the organophosphorus anticholinesterases which kill rapidly when traces make contact with the skin should serve as a salutary reminder of the possibility of transdermal toxicity. In 1954, 105 people died in France after skin treatment with Stalinon, a topical antiseptic containing diethyl tin di-iodide, which had been manufactured in a pharmacy. In this instance, the formulation was altered after initial clinical studies had been completed. This is one of the reasons why regulatory authorities now insist that trials be conducted on the formulation that will ultimately be marketed.

Severe neurotoxicity has occurred in young infants from the absorption of hexachlorophane through skin. This antiseptic, which has the advantage of retaining its activity in the presence of soap, was widely used in creams, soaps and cleansing lotions sold to the general public for many years before the dangers were recognized. Pituitary repression and frank hypercorticism can be caused by prolonged application of high-potency corticosteroid preparations to the skin, especially if occlusive dressings are employed. This is more likely to happen when steroids are applied to areas where absorption will be greatest, such as thin skin, folded skin or raw surfaces. Infants are at greatest risk.

The extensive toxicity testing that is now mandatory for all new drugs and formulations is based in no small measure on what has been learned from the incidents that are described in this chapter. As awareness of past problems may alert the practitioner to the possibility of what could transpire in the future, the remainder of this chapter deals with some of the serious adverse drug reactions that have affected various organs or systems.

BLOOD DYSCRASIAS

Blood dyscrasias constitute the most serious of all adverse drug reactions since more than one quarter result in death. Figures released by the UK Committee on Safety of Medicines disclose that nearly one fifth of all voluntarily reported fatal adverse drug reactions are attributable to this cause alone. If this fact causes surprise, it can be ascribed to the rarity with which drug-induced blood dyscrasias actually occur. The CSM has received an average of just over two hundred reports of these each year over the past 20 years, which means few physicians will have observed this type of adverse reaction.

The probability of elderly patients developing a blood dyscrasia is slightly greater than for younger patients, but 40 per cent of the former die. This might possibly be because elderly patients are more likely to develop aplastic anaemia, which has a mortality rate in the order of 70–80 per cent. The first signs of it usually involve bleeding, and the majority of patients develop pancytopaenia. The increased availability of bone marrow transplantation may well lower the mortality rate for younger patients in the future.

The Registry on Blood Dyscrasias was inaugurated for the American Medical Association in 1954 by its Council on Pharmacy and Chemistry, later being expanded to include all adverse reactions. This compilation of reports had been instituted following disquiet over repeated assertions of an association between chloramphenicol and deaths from aplastic anaemia. This orally active, broad-spectrum antibiotic had come into general use in 1948. Over eight million patients received it before its ability to induce a delayed aplastic anaemia was first recognized. The incidence of this has since been put at somewhere between 1 in 20 000 and 1 in 100 000. Because of the clinical value of chloramphenicol at that time, the US Food and Drug Administration did not call for its withdrawal; instead, warnings were printed on package inserts.

In 1962, the Registry of Blood Dyscrasias reported several cases of agranulocytosis and aplastic anaemia amongst twelve patients who had developed haematological abnormalities associated with consumption of phenylbutazone, an antirheumatic drug. Although numerous clinical reports confirmed the hazard, phenylbutazone remained freely prescribable in the United Kingdom until the Committee on Safety of Medicines intervened in 1984 to restrict its use to the treatment of ankylosing spondylitis. Figures released at that time indicated that phenylbutazone had probably been responsible for over three hundred deaths during the preceding 20 years. This has to be considered in conjunction with the fact that at the height of its popularity in the late 1960s, British general practitioners were annually issuing more than three and a half million prescriptions for phenylbutazone. The related oxyphenbutazone was more toxic and had its licence revoked. A few months later, feprazone was withdrawn from the market by its manufacturer because the cost of generating the extra safety data required for the continuation of its product licence would have been prohibitive. Deaths from aplastic anaemia have followed the use of other antirheumatic agents, notably gold salts and penicillamine, as well as in elderly patients given co-trimoxazole, a valuable combination preparation containing sulphamethoxazole and trimethoprim.

Drug-induced agranulocytosis is frequently due to hypersensitivity, but in the case of phenothiazine tranquillizers such as chlorpromazine it appears to be due to a direct damaging action on the developing granulocytes in the bone marrow. Typically, this occurs only when cumulative doses in excess of 10 g have been administered. Withdrawal of the drug usually results in recovery of the bone marrow unless the patient succumbs to overwhelming infection.

The nature of the bone marrow depression induced by cytotoxic drugs used in cancer chemotherapy is well recognized. These drugs attack all rapidly dividing cells in the body, this being an extension of their therapeutic action. This is predictable and dose-related in a relatively clear-cut manner, although the severity of the reaction and its temporal characteristics may vary somewhat. The most dramatic effects on the blood are first seen on the platelets and leucocytes as these have a life span of only one week. Both the bone marrow depression and the anaemia that can also occur during cancer chemotherapy are reversible if therapy is promptly withdrawn. To do this in time requires constant blood monitoring.

HEPATOXICITY

The Committee on Safety of Medicines views drug-induced liver disease with concern since 14 per cent of all cases that have been reported had a fatal outcome. In 1964, the register of adverse drug reactions compiled by its predecessor, the Committee on Safety of Drugs, identified a very high incidence of jaundice amongst the relatively small number of patients taking the newly introduced coronary dilator known as benziodarone. It was found that adverse effects were produced in up to 10 per cent of patients who took the drug for more than 2 months. This resulted in the voluntary withdrawal of the drug by its manufacturer. The chemically related antiarrhythmic agent amiodarone has also caused liver damage in some patients. Tienilic acid (ticrynafen) also had to be withdrawn from the market, in 1980, after the deaths in the United States of 24 patients from severe liver damage associated with its use. It was an analogue of the diuretic agent ethacrynic acid and possessed uricosuric properties. Benoxaprofen has already been discussed; one of its severe toxic actions was hepatoxicity. This was particularly prevalent in elderly patients in whom drug accumulation occurred, but it is noteworthy that the incidence of drug-induced liver disease generally increases with age.

Over three hundred drugs have been reported to the Committee on Safety of Medicines as having been associated with the occurrence of jaundice, although this only accounts for 3 per cent of all serious adverse reactions. The most frequently cited drugs are halothane and chlorpromazine. These are discussed in the next chapter since their hepatoxicity is believed to be of an allergic nature.

Several commonly used drugs have been found capable of exerting a direct toxic action on the liver. One of these is paracetamol (acetaminophen), which is converted in the liver mainly to glucuronide or sulphate conjugates. However, traces are oxidized to the toxic quinone form and this is detoxified by reacting with free glutathione. If an overdose of paracetamol is taken, the liver cells are damaged by excess of paracetamol quinone. Aspirin and teracyclines can also be hepatoxic if high doses are administered for prolonged periods.

Before the dosage of oestogens in oral contraceptives was reduced, the occasional appearance of jaundice in the first few cycles of medication was not uncommon. This was due to increased levels in plasma of bilirubin as a consequence of oestrogen-induced cholestasis. This phenomenon is still seen in patients taking anabolic steroids.

GASTRO-INTESTINAL TOXICITY

Patients receiving antirheumatic analgesics frequently report gastro-intestinal symptoms. In 1938, the newly developed gastroscope revealed that aspirin caused bleeding and sometimes even ulceration. The soluble calcium salt was less harmful, yet it was many years before the greater safety of this salt was generally acknowledged. However, even the use of soluble aspirin does not eliminate the problem. In 1955, it was shown that one third of a large group of patients admitted to hospital with severe gastro-intestinal bleeding had consumed aspirin or another salicylate within the previous six hours. However, in 1974 Levy carried out an extensive investigation as part of the Boston Collaborative Drug Surveillance Program and established that the incidence of gastric bleeding which required hospital admission was a mere 15 out of every 100 000 members of the adult population. He also found that the incidence of uncomplicated gastric ulcer was about two thirds of this figure. Since then, many other drugs have been implicated as a cause of gastric bleeding, although this has rarely been found to require hospitalization. Phenylbutazone, oxyphenbutazone and indomethacin can cause gastric ulceration. In 1973, British and American doctors reported a high incidence of severe and sometimes fatal pseudomembranous colitis in patients receiving either lincomycin or clindamycin. This was disputed by the manufacturer of these chemically similar antibiotics, who argued that the incidence of this was only 1 case per 100 000 patients, as based on reports received over a 10-year period. Several investigators supported the manufacturer in this controversy, but the commercial prospects for the drugs were severely curtailed. Further investigations confirmed that the colitis was caused by a toxin released from *Clostridium difficile*, an antibiotic-resistant anaerobe. The risk of this was greatest in middle-aged and elderly women, particularly after surgery. It occurred very rarely with other antibiotics, including ampicillin.

RENAL TOXICITY

A variety of drugs are known to damage the proximal renal tubules, with excessive blood levels of antibiotics constituting the commonest cause. Amphotericin, the most valuable systemic antifungal agent available, is highly nephrotoxic. The toxicity is dose-related and some degree of renal impairment occurs in most patients. Dose-related, reversible renal failure has been frequently observed in patients receiving aminoglycoside antibiotics, although ototoxicity remains the principal hazard. Problems usually occur after treatment has lasted about one week. Gentamicin, the aminoglycoside most widely prescribed for treatment of serious infections, appears to be more nephrotoxic than netilmicin. Cephaloridine, a first generation cephalosporin, has been superseded on account of its troublesome nephrotoxicity.

Besides antibiotics, particular concern has been occasioned by the nephrotoxicity of certain heavy metals and organometallic compounds. Notable examples are mercurial diuretics, the cytotoxic agent cisplatin, and gold salts used in the treatment of rheumatic disease, although the latter are more likely to affect the kidney indirectly as a consequence of hypersensitivity reactions.

In 1953, attention was drawn to the occurrence of renal papillary necrosis associated with excessive consumption of compound analgesic mixtures. This was first recognized amongst female workers in the Swiss watch industry, whose work frequently induced eye strain and headache. Suspicion fell on phenacetin, the ingredient common to most analgesic preparations. Because the nephropathy was only seen in patients who had consumed analgesics daily for several years, and the incidence was therefore low, regulatory authorities were reluctant to call for withdrawal of the drug. Nevertheless, around 10 per cent of cases of chronic renal failure were due to analgesic nephropathy. By 1974, the apparently safer phenacetin metabolite known as paracetamol had become so well established that there was little public outcry when phenacetin preparations became available only on prescription in the United Kingdom. Since then, it has been suggested that phenacetin was unfairly singled out as the sole causative factor in renal necrosis, abuse of all compound analgesic preparations being held responsible.

NEUROTOXICITY

The most valuable of all antituberculous drugs is isoniazid, introduced in 1952. High doses were found to cause peripheral neuritis by disrupting pyridoxine (vitamin B_6) metabolism in nerve tissue. Fortunately, this could be treated by administration of the vitamin. It is now known that the drug reacts chemically with pyridoxal phosphate to form a hydrazone which inhibits pyridoxal phosphate kinase. As with other forms of isoniazid toxicity, the occurrence is more frequent amongst slow acetylators.

Clioquinol (iodochlorhydroxyquin) was first marketed in 1934 and went on to become a popular oral antidiarrhoeal remedy throughout the world. In 1970, the Japanese Ministry of Health and Welfare banned the sale of the drug following its alleged involvement in an ongoing epidemic involving possibly ten thousand cases of subacute myelo-optic neuropathy (SMON). Patients who had consumed large doses of clioquinol for at least two weeks were presenting with damage to the optic nerve, spinal cord and peripheral nerves. Besides mild to severe visual disturbance, they were experiencing difficulty in walking due to weakness and stiffness of their legs. For many, the disease process turned out to be reversible, but those most severely affected were permanently blinded. As soon as the government action was taken the epidemic ceased. Few cases were ever reported outside Japan despite the worldwide use of clioquinol as a prophylactic against traveller's diarrhoea. Animal tests have consistently failed to reproduce the effects seen in humans.

OCULAR TOXICITY

The most troublesome toxic manifestation that Paul Ehrlich had to overcome when developing organic arsenicals as therapeutic agents was their ability to damage the optic and aural nerves. Shortly after the death of Ehrlich, Morgenroth's ethylhydrocupreine was sold on the Continent and in the United States as the first antibacterial chemotherapeutic agent. This quinine analogue had similar neurotoxic properties to the arsenicals, causing blindness when high blood levels were achieved in patients being treated for pneumonia. After being reformulated as an oral suspension of the almost insoluble free base in order to avoid high peak plasma levels by slowing the rate of absorption from the gut, it remained on the market until the introduction of sulphapyridine in 1938. The problem of neurotoxicity has continued to reappear with several modern chemotherapeutic agents, sometimes limiting their effectiveness.

Ethambutol, a valuable antituberculous drug, was introduced in 1961 and was found to be relatively free of side-effects so long as unnecessarily high dosage was avoided. Damage to the optic nerve occurred in many patients receiving excessive amounts of the drug, hence the dose level had to be kept under control, especially if the ability to excrete it was reduced by renal impairment. Fortunately, most patients usually became aware of visual disturbance before the retrobulbar neuritis resulted in irreversible damage. This meant that when ethambutol was prescribed, patients had to be warned to report any visual disturbance immediately. Early investigations into the neurotoxicity revealed that the dextrorotatory optical isomer was somewhat less troublesome, since when formulations have incorporated this rather than the racemic mixture formerly used.

In the early 1960s, rheumatologists began to use massive doses of the antimalarial chloroquine in patients severely crippled with arthritis. It was eventually found that after prolonged medication for perhaps two or more years, irreversible retinal changes occurred, sometimes causing blindness. These have been attributed to the strong affinity of the drug for melanin in the choroidal and retinal epithelia. Phenothiazine tranquillizers can also cause retinal damage by binding to melanin, the worst offender being thioridazine when large doses are administered chronically. The visual disturbances seen in most patients suffering from overdosing with digoxin are thought to be due to an effect of the drug on photoreceptors.

The hypocholesterolaemic synthetic oestrogen analogue triparanol was available for only a short time before being withdrawn from the American market in 1962 because it induced cataracts in several patients. There has been speculation that this could have been caused by permeability changes arising from either the effects of the drug on lipid metabolism or on lens membrane structure. Triparanol was not the first drug to produce cataracts. During the 1930s, dinitrophenol was administered as a slimming drug which acted by raising metabolic rate (due to uncoupling of oxidative phosphorylation). In addition to inducing cataract formation, it could cause death from agranulocytosis. Cataract formation is also a hazard when high dosage corticosteroid therapy is continued for upwards of one year. The topical application of the same compounds can raise intra-ocular pressure.

OTOTOXICITY

Aminoglycoside antibiotics are toxic to the outer hair cells of the cochlear epithelium. The degree of ototoxicity varies with different aminoglycosides, being most severe with neomycin. As the damage is dose-related and affects the perception of high frequency sounds first, frequent plasma monitoring and audiometric testing can do much to prevent deafness. Some aminoglycosides, such as streptomycin, affect the vestibular apparatus before hearing loss occurs, leading to balance problems. Damage by gentamicin is largely vestibular, whilst netilmicin seems to be both less ototoxic and less nephrotoxic. Cisplatin is another nephrotoxic drug which causes ototoxicity.

Rheumatic patients on high-dose aspirin therapy frequently experience reversible tinnitus, an unpleasant ringing sensation. This may occasionally progress to cause temporary deafness.

CARDIOVASCULAR TOXICITY

Only 11 weeks after Simpson introduced chloroform in 1847, a healthy young woman in Newcastle died from sudden cardiac failure whilst having a toe-nail removed. Her death was to prove typical insofar as she was undergoing only mild anaesthesia for a trivial operation. Occasional deaths of this nature continued to occur and were attributed to cardiac arrest caused by the phase of excitement that was most noticeable during light anaesthesia. In 1864, the Royal Medico-Chirurgical Society established a committee to inquire into 123 such reports. It conducted experiments which appeared to confirm the view that chloroform was cardiotoxic. It also suggested that mixtures of chloroform and ether might be safer. Despite several further commissions of inquiry by a variety of authorities, the number of deaths annually attributed to chloroform in England remained around 60 during the last few years of the nineteenth century. There was not even agreement amongst physiologists as to the precise nature of the action of chloroform on the human heart. Experiments on different animal species gave conflicting results. It was not until 1911 that A. Goodman Levy, acting on a suggestion from Professor Arthur Cushny of the University of London, demonstrated that chloroform could induce ventricular fibrillation in lightly anaesthetized cats. The same effect could be achieved by intravenous injection of adrenaline, but once again only in lightly anaesthetized animals. A quarter of a century was to pass before Ralph Waters at the University of Wisconsin demonstrated that chloroform sensitized the heart to the action of adrenaline. Shortly before this, he had been instrumental in securing the introduction of cyclopropane, one of several safer drugs that were ultimately to displace chloroform in modern anaesthetic practice.

Pressurized isoprenaline aerosols for the relief of asthmatic attacks were introduced in the United Kingdom in 1961, but four years passed before any suspicions arose concerning the safety of these convenient products. One report of eight unexpected sudden deaths amongst users of these was followed by another by Dr Gerald Corner drawing attention to an increasing death rate in asthmatic children in the Bristol area. By examining the Registrar General's statistics, he then confirmed that this had been the national pattern since 1960 amongst children throughout England and Wales. Detailed epidemiological studies were immediately initiated with a view to establishing whether the mortality figures paralleled the sales of isoprenaline aerosols. In June 1967, without delaying until this study could be completed, the UK Committee on Safety of Drugs issued a warning to doctors and restricted the availability of the aerosols to prescription only. Warnings of the danger of excessive use were printed on aerosol labels. These steps proved highly effective, and the death rate from asthma declined quickly to reach pre-1960 levels. The epidemiological reports which appeared a year or two later never proved conclusively that the aerosols were directly responsible, but circumstantial evidence suggested that around three thousand young teenagers had died from excessive use of these devices. There was much speculation as to the circumstances in which isoprenaline could prove cardiotoxic, but no precise mechanism was ever established.

The clinical value of the anthracycline antibiotic doxorubicin in cancer chemotherapy has been severely limited by its cardiotoxicity. Cumulative doses in excess of 200 mg/m^2 have produced congestive heart failure. Its recently introduced isomer known as epirubicin is safer, possibly because it can be detoxified in the liver to form a glucuronide. Heart failure is only produced after cumulative doses in excess of 1000 mg/m^2. The severity of other side-effects is also reduced in comparison with doxorubicin.

The cardiotoxicity of digitalis glycosides has been referred to in the first chapter. Much of the problem stems from the difficulty of distinguishing the toxic actions of these compounds from those of the disease process being treated. Since digoxin, the most frequently used cardiac glycoside, has to be administered at around 60 per cent of the toxic dose level in order to achieve therapeutic benefit, it is hardly surprising that from one tenth to one quarter of all patients receiving it suffer toxic effects.

In 1976, the antimetabolite azaribine was withdrawn from the American market after it was reported to be capable of causing severe thrombosis. It had been prescribed for fungal infection of the skin and psoriasis. This was not the first time concern about this type of toxicity had arisen, for it had previously been encountered in women taking oral contraceptives.

The first oral contraceptive was granted approval for marketing in the United States in May 1960 after the Food and Drug Administration had examined the records of 1600 women who had participated in clinical trials initiated four years earlier. The oral contraceptive was marketed in the United Kingdom a few months later. The timing is significant, for it was in November 1961 that the teratogenicity of thalidomide was first revealed. It was inevitable that in the ensuing hysteria about drug safety, the oral contraceptive would come under close scrutiny. Firstly, the risk–benefit ratio had to be considered in view of the drug having no therapeutic value. Other than pressing social reasons, all that could be adduced in its favour in this connection was the argument that pregnancy itself constituted a tangible risk. As the maternal mortality rate in some countries has halved since then, this is now less significant than it once was. The second point of concern was over the fact that many women would take oral contraceptives for a high proportion of their life span. One answer given to those concerned about the long-term consequences of this was that in primitive times women existed in a near constant state of pregnancy where the hormonal effects were not unlike that produced by oral contraceptives. Such a flippant assertion convinced few critics.

A high proportion of women reported side-effects and ceased taking oral contraceptives. Whilst changes in fluid balance and other drug-related problems were common with the early high-dosage formulations, it seems likely that psychosomatic responses were encouraged by adverse publicity in the media. Such reports were an understandable reaction to the thalidomide episode. However, particular concern centred on the possibility of an increased risk of thromboembolic disease in otherwise healthy women. This was considered to be related to the effect of oestrogens on blood clotting. In an attempt to settle the matter, the Food and Drug Administration established an expert committee which examined the medical records of 350 women who had developed thromboembolic disease whilst taking oral contraceptives. Although their mortality rate was slightly higher than in a comparative group of non-users, the results were not statistically significant.

In 1965, the Committee on Safety of Drugs published the report on its own investigations into the deaths of women who had been taking oral contraceptives. This concluded that although the number of reported cases of fatal pulmonary embolism amongst users was similar to that of the general population, the matter should be investigated further. There were grounds for suspecting that a high degree of underreporting of adverse reactions existed, since the reporting rate from clinics of the Family Planning Association was two and a half times greater than that from general practitioners. Accordingly, the Committee on Safety of Drugs set up a committee of inquiry, which stated, in 1968, that there was a slightly increased risk of thromboembolism (estimated to be an excess of 3 deaths per 100 000 users) amongst women taking oral contraceptives, a conclusion that was in agreement with independent reports published the previous year by the Royal College of General Practitioners and by Doll and Vessey of the Medical Research Council's epidemiology unit at Oxford. Further analysis of the information available to the Committee on Safety of Drugs and also the regulatory authorities of Sweden and Denmark disclosed that the risk of deep vein thrombosis or pulmonary embolism was related to the amount of synthetic oestrogen in oral contraceptives. Accordingly, it was recommended that in future only low levels of oestrogen should be incorporated.

In 1973, investigators from the Boston Collaborative Drug Surveillance Program claimed that although the incidence of idiopathic venous thrombosis was 6 per 100 000 amongst non-users, it rose elevenfold to 66 per 100 000 amongst women taking oral contraceptives. The risk factor was also shown to be greater amongst women who smoked. British concern over the safety of oral contraceptives was heightened by a 1975 report from the Royal College of General Practitioners. This was based on the analysis of data collated by a large number of family doctors. It confirmed the fears aroused by the Boston study by claiming that the death rate amongst women who had taken oral contraceptives continuously for five years was ten times that of non-users. Women over the age of 35 were at greatest risk. This report aroused some controversy that was further fuelled when the Medical Research Council team claimed that there appeared to be an increased incidence of myocardial infarction amongst young women taking oral contraceptives. Subsequently, the Boston workers confirmed that this increased risk of myocardial infarction applied mainly to women who smoked. Numerous studies which have appeared since the publication of those described here tend to confirm the earlier findings.

PULMONARY TOXICITY

Aminorex, a cyclic analogue of amphetamine, was marketed in Europe in the late 1960s for use as an anorectic. In 1969, the UK Committee on Safety of Drugs declined to release it for clinical trial because of reports that the drug could cause pulmonary hypertension. Subsequent investigations in Europe revealed that the incidence of pulmonary hypertension had risen by one fifth since aminorex had been introduced.

Bleomycin is an antitumour antibiotic used mainly in the treatment of squamous cell carcinomas. Although it is one of the few anticancer agents which does not cause bone marrow depression, one patient in every hundred dies from pulmonary fibrosis. This is progressive, and occurs mainly in elderly patients and in those who have received cumulative doses in excess of 300 mg. About one patient in every ten develops pneumonitis or some degree of pulmonary fibrosis. Pulmonary damage may also be caused by long-term therapy with nitrofurantoin. This not infrequent occurrence is characterized by an acute pleural effusion which is reversed on cessation of therapy.

TERATOGENICITY

The greatest drug disasters have followed in the wake of the pharmacological revolution of the 1950s and 1960s. Paradoxically, the most notorious incident involved a drug that was widely promoted on account of its alleged safety as a non-barbiturate hypnotic-sedative. Thalidomide was introduced on the German market in 1956. In November 1961, Dr W. Lenz, a Hamburg paediatrician, reported a large increase in the number of infants with phocomelia attending ten clinics in North Germany. This was hitherto one of the rarest malformations known, without any cases having been seen in these clinics in the decade prior to 1959. Yet the number of cases had escalated from ten that year to 477 in 1961. The increase was attributed to the taking of thalidomide by mothers during the first trimester of their pregnancies. The drug had never been tested for teratogenicity. The report of thalidomide teratogenicity was immediately picked up and sensationalized by a German newspaper, forcing the manufacturer to withdraw the drug. By then, three thousand deformed babies had already been born in Germany and about twice that number elsewhere. The United States was spared from the thalidomide disaster because the Food and Drug Administration had not approved a new drug application, having been dissatisfied with the limited safety data that had been submitted.

Around the same time that thalidomide was introduced, a compound preparation containing the antihistamine doxylamine, the anticholinergic dicyclomine and pyridoxine, a B complex vitamin, was marketed as 'Bendectin' in the United States and as 'Debendox' in the United Kingdom. It was recommended for the prevention of nausea in early pregnancy. Following the revelation of the teratogenicity of thalidomide, all drugs frequently taken by pregnant women became the subject of much scrutiny. Repeatedly, this widely consumed preparation was singled out for strong criticism, it being alleged to have caused foetal malformation. As a result of a sustained campaign of hostile publicity, its manufacturer felt compelled to withdraw the preparation from the market in 1983. In the United States, a legal case against the manufacturer was settled out of court on the grounds that the cost of defending it would have been greater than the sum claimed as compensation by mothers who had taken the drug.

The problem with 'Debendox'/'Bendectin' was that as an estimated 33 million women had taken it, and the incidence of spontaneous birth defects amongst non-users lies somewhere between 2 and 7 per cent, it was inevitable that a substantial number of women who spontaneously gave birth to a deformed child would blame the drug. The outcome of several scientific inquiries has made it clear that if there was any increased incidence of birth defects, it was very low and of questionable significance. Because of the uncertainty surrounding the size of any slightly increased level of incidence, it is no simple matter to prove that the drug might be responsible in a given case. The matter is being contested in the courts.

The vitamin A analogues isotretoin and etretinate are used in the oral treatment of severe acne and psoriasis, respectively. Both are teratogenic, hence female patients must adopt contraceptive measures. In the case of etretinate, these must continue for one year after cessation of treatment since the drug persists in the body for several months.

CARCINOGENICITY

One of the greatest concerns of those who are involved with the introduction of new drugs is that the appearance of a serious adverse reaction, especially a cancer, may not occur until some years after a patient takes the drug. Basically, there appear to be two ways in which a drug may induce cancer. The first involves the formation of a trace metabolite which has the ability to function as an electrophile and thus cause genotoxicity by attacking a nucleophilic centre in DNA or some other vital cell component. Occasionally, even the unchanged drug has this activity. The second method of inducing cancer is indirect, or epigenetic. This could be elaborated by hormonal changes, oncoviral stimulation, or through altering present or future susceptibility to either an environmental carcinogen or a genotoxic drug.

The first clinically effective beta-adrenoceptor antagonist (i.e. beta-blocker) was pronethalol, introduced in 1962. Shortly after completion of large-scale clinical trials, it was found to induce cancer of the thymus in mice. It was immediately replaced by the more potent and less toxic analogue known as propranolol. This was non-carcinogenic.

The results were published in 1965 of an analysis of the records of 1107 patients who, during the period of 1930–1952, had been injected with a colloidal solution of thorium dioxide, an X-ray contrast medium used in liver or spleen arteriography. Local granulomas were found to have been detected in 81 patients, whilst a characteristic haemangioendothelioma was seen in 22 others. In addition, 42 cases of cirrhosis of the liver and 16 deaths from blood dyscrasias had been reported.

In 1970, A.L. Herbst reported seven cases of vaginal adenocarcinoma in American women aged 15–22. Previously this had usually been seen in post-menopausal patients. The following year, Herbst published the findings of a case-control study which revealed an association between the development of clear cell adenocarcinoma of the vagina in young women and their exposure *in utero* to stilboestrol (diethylstilbestrol) taken by their mothers for the maintenance of pregnancy. His conclusion was confirmed by several other workers and by reference to cancer registries.

Cancer of the cervix was also found to be associated with uterine exposure to synthetic oestrogens. The related compounds hexoestrol and dienoestrol were also implicated in both types of malignancy. In the light of these observations, the Food and Drug Administration banned the use of stilboestrol and dienoestrol during pregnancy. It was fortunate that the incidence of cancer was low since it has been estimated that in America synthetic oestrogens had been administered to somewhere between 500 000 and 2 000 000 pregnant women between the years 1945 and 1955. However, there have recently been reports of testicular cancer amongst men in their late 20s whose mothers had taken stilboestrol during pregnancy.

The high-potency progestogen chlormadinone acetate was withdrawn by its manufacturer in 1970 after experiments on beagle dogs revealed an increase in the proportion of small nodules in their breasts. As the drug had been incorporated in oral contraceptive formulations, this observation was considered to be ominous. Similar action had later to be taken with the close structural analogue megestrol acetate, for the same reason. In the case of medroxyprogesterone acetate, not only was the proportion of breast nodules increased, but some were found to be malignant and metastases were present. However, there was at that time no convincing evidence that high-potency progestogens caused breast cancer in humans.

The organizers of the Boston Collaborative Drug Surveillance Program examined the case histories of 800 patients with newly diagnosed tumours. Their drug histories were compared with those of 3443 hospitalized patients who were free from cancer. Patterns of regular drug usage in both groups were found to be similar. From this study it emerged that less than 1 per cent of the malignancies could be correlated with regular drug usage, but that almost half of those in males could be due to smoking. The only drug which came under suspicion was reserpine. In 1974, evidence was presented from this multiple-comparison study to suggest that the risk factor for breast cancer was 3.5 times greater for women taking reserpine to control hypertension. Several independent investigators supported this conclusion, but it was disputed by others.

In 1983, Professor Malcolm Pike published the results of his examination of the medical records of 314 women in Los Angeles under the age of 37 who had developed breast cancer. The findings suggested that the long-term use of oestrogen-progestogen oral contraceptives before the age of 25 could have been associated with an increased risk of breast cancer, but only in those patients who had taken preparations containing very high potency progestogens. The available data did not establish a direct causal linkage between the disease and such preparations. Two months later, a British study by Professor Vessey at Oxford adduced evidence supporting Pike's findings and also an association between the long-term use of oral contraceptives and an increased incidence of cervical cancer. Again, it was not possible to establish a direct causal relationship, because other factors may have been involved. For example, sexual behaviour in women taking oral contraceptives may differ from that of those who use other forms of contraception. In the light of these two reports, the Committee on Safety of Medicines recommended that women should be switched to alternative preparations containing low doses of oestrogens and less potent progestogens. It also endorsed Professor Vessey's recommendation that since all the cases of invasive cancer detected in his study through cervical cytology responded successfully to treatment, regular cytological examination was advisable for long-term users of oral contraceptives.

However, in 1984 a report from the Cancer and Steroid Hormone Group (CASH) claimed that its findings did not establish a significant association between the use of contraceptives containing high-potency progestogens before the age of 25, or before the first normal pregnancy, and the development of breast cancer before the age of 37. In 1983, a CASH study had suggested that oral contraceptive use offered some protection against ovarian and endometrial cancer. Two years later, the preliminary results from a study organized by the UK Royal College of General Practitioners supported this view.

These conflicting results about the safety of oral contraceptives highlight the problem faced by investigators seeking to prove that a drug has been the cause of a low-incidence adverse reaction which also occurs spontaneously in the untreated population at large. Even when such an association is firmly established, it is difficult to prove this is one of cause and effect, unless the withdrawal and reintroduction of the drug influences the incidence of the adverse reaction.

V

Allergic Reactions to Drugs

The pattern of adverse drug reactions was altered by the introduction of synthetic drugs which had undergone extensive selection procedures that ensured low levels of toxicity. As the incidence of untoward toxic effects steadily declined, the enhanced safety enabled more drugs to be administered repeatedly. Consequently, allergic reactions to drugs occurred with increasing frequency. De Weck and Bundgaard, the editors of a recent book on allergic reactions to drugs, have suggested these might now account for one third or even one half of all adverse drug reactions. Others would dispute this. Whatever their incidence, these reactions remain largely unpredictable insofar as no reliable animal models exist. This prevents pharmaceutical companies from developing safer alternatives before clinical trials begin. On several occasions recently, otherwise promising drugs have had to be withdrawn from the market due to an unacceptable incidence of hypersensitivity reactions that were entirely unforeseen.

The commonest drug allergies take the form of skin reactions. During the 1970s, when ampicillin was so widely prescribed for the most trivial of reasons, it achieved the notorious distinction of being the single compound responsible for the greatest number of adverse drug reactions. Fortunately, these were mainly skin rashes. Severe or fatal immune reactions to drugs remain rare, being more likely to appear when medicines are administered intravenously rather than intramuscularly or by mouth. Patients with a history of allergy or asthma are, understandably, more at risk of becoming hypersensitive to drugs than are others.

The pioneering studies of Karl Landsteiner in the 1930s established that low molecular weight compounds, when covalently bound to proteins, confer antigenicity. Drugs can do likewise if a stable chemical bond with a protein or other high molecular weight carrier can be formed. The bond usually has to be strong enough to resist metabolic degradation. Drug–protein bonding can occur either *in vivo* or as a result of exposure to a protein during the manufacturing process if the drug is obtained from natural sources. The latter phenomenon was observed in 1967 when scientists at the Beecham Research Laboratories detected trace amounts of penicilloyl-protein formed by reaction between benzylpenicillenic acid and protein in commercial samples. This acid was formed by chemical rearrangement of benzylpenicillin in acid solution; it should not be confused with the penicilloic acid, the major breakdown product formed on hydrolysis of the beta-lactam ring. Penicillenic acids, in general, are powerful acylating agents. Since the penicilloyl-protein detected by the

Beecham researchers was strongly antigenic, it was held to be at least in part responsible for the frequently occurring hypersensitivity reactions to the antibiotic. Others disputed the claim because they detected far smaller amounts of this antigenic contaminant. They argued that hypersensitivity reactions were due not to it, but rather to an autologous penicilloyl-protein formed *in vivo* by reaction of the penicillenic acid with an endogenous protein. As the penicilloyl residue acted as the immunogenic determinant, this would have been less antigenic than heavily penicilloylated heterologous protein formed in the presence of high concentrations of penicillin during fermentation. Furthermore, little penicillenic acid is produced *in vivo*. Ten years passed before the issue was resolved by the development of a sensitive radioimmuno-assay method for the quantitative determination of penicilloyl-protein present in commercial products. This confirmed that the amount varied from one to three parts per million in different samples of a variety of natural and semi-synthetic penicillin formulations. Almost all samples contain small amounts of penicillenic acid as this forms most readily under the acid fermentation conditions that must be maintained to minimize hydrolysis of the beta-lactam ring. Experiments on animals indicated that whilst highly purified ampicillin could only produce a very weak antibody response, heavily con-taminated penicillins induced a considerable degree of antibody production. Improved purification techniques over the years have reduced the incidence of penicillin hypersensitivity.

Amino-penicillins may undergo self-polymerization through nucleophilic attack by the non-ionized amino group of one molecule on the beta-lactam ring of another. The process can be repeated a few times to form polymers which have been shown to be antigenic in animals. Ampicillin is more susceptible to polymerization than other penicillins, so this may account for the unusually high incidence of hypersensitivity reactions.

Most drugs which cause hypersensitivity reactions are not contaminated with antigenic material in the manner that has been seen with penicillins. It is now generally accepted that *in vivo* many drugs generate minute amounts of immunogenic metabolites or degradation products which can bind covalently with endogenous proteins or other autologous carriers. The resulting combina-tion is antigenic. For example, metabolic hydroxylation of an aromatic ring will permit phenols and aniline derivatives to form highly reactive quinones or quinoneimines that can attack non-ionized amino groups on proteins. An alternative activation process could involve oxidation of ethylenic bonds to produce epoxides. It is pertinent to note that that these reactive intermediates are also believed to be mutagenic or carcinogenic through their ability to react with DNA. It should be noted that most of the drugs mentioned in this chapter, or their principal metabolites, are organic acids.

It is rarely possible to identify the homologous carrier to which an im-munogenic drug has been able to bind. One notable exception to this is the identification of the penicilloylated red cell membrane as a site of attack of I_gG

antibodies to penicillin, resulting in haemolytic anaemia. This only occurs after prolonged, high-dosage penicillin therapy, and is by no means a typical penicillin hypersensitivity reaction.

ANAPHYLACTIC REACTIONS

These reactions vary from mild urticaria to fatal anaphylactic shock, being characterized by the rapidity with which they occur after exposure of a sensitized individual to the provocative agent. Anaphylactic shock involves not only hypotension, but also bronchoconstriction and congestion of mucosal surfaces of the lungs and other organs. The speed with which these effects occur is a consequence of the strong binding of antibodies (i.e. reagins) to the membranes of mast cells in the tissues and basophils in the blood. The antibodies have been characterized as the I_gE type of immunoglobulins. Their binding to the cell membranes somehow triggers the immediate release into the circulation of such mediators as histamine and prostaglandins, as well as related inflammatory substances. The ensuing increase in capillary permeability allows escape of protein from the blood to produce oedema in surrounding tissues.

Since shortly after Ehrlich and von Behring introduced diphtheria antitoxin at the end of last century, it has been recognized that there is a risk of anaphylactic reactions after the parenteral administration of antitoxins obtained from animal serum. Vaccines and toxoids prepared from microorganisms rarely cause allergic reactions, but viral vaccines grown in eggs can occasionally give trouble. Today, the problem of anaphylaxis is seen at its worst with the penicillins and cephalosporins. Whilst they usually produce only urticaria, on rare occasions severe and even lethal anaphylactic shock may occur within minutes of administration.

Allergic reactions to modified gelatin or dextran infusions given as plasma expanders in the treatment of certain types of shock occur infrequently. However, when lower molecular weight dextrans form a colloidal complex with ferric hydroxide, known as iron dextran injection, there is a greater risk of severe or even fatal anaphylactic shock if this is given by infusion rather than as a course of intramuscular injections. In situations where this route of administration is unavoidable, a small test dose must first be tried. If no reaction occurs within half an hour, a slow infusion may proceed.

Zomepirac, a non-steroidal anti-inflammatory agent, was voluntarily withdrawn by its manufacturer in 1983 after being on the market for three years. The drug was estimated to have been taken by twenty million patients around the world. There had been five reports of fatal anaphylactoid reactions in the United States, but in the United Kingdom no deaths had occurred. Two deaths were in patients said to be hypersensitive to aspirin, but the issue of aspirin hypersensitivity itself is difficult to resolve. The adverse reaction to aspirin often found amongst asthmatics is best described as intolerance (see page 42).

However, aspirin allergy does exist; whether it or intolerance is the main cause of adverse reactions is debatable.

Anaphylactic reactions have been caused by Cremophor EL, a surfactant used to render intravenous anaesthetics such as propanidid and alphaxalone-alphadolone acetate water-soluble. In 1983, the former product was withdrawn by its manufacturer, and the latter likewise a few months after.

CYTOTOXIC REACTIONS

Some hypersensitivity reactions to drugs involve cell destruction. Blood cells are frequently affected, but renal and hepatic cells may also be damaged. The first cytotoxic hypersensitivity reaction to be recognized was that involving penicilloylated red cell membranes after prolonged, high-dosage penicillin therapy. A different type of haemolytic reaction involves binding of a pre-formed drug-antibody complex on to the red cell surface together with complement components. This type of reaction has been detected with aminosalicylic acid, insulin, melphalan, phenacetin, quinine, quinidine, rifampicin, stibophen and sulphonylurea hypoglycaemic agents.

Cytotoxic reactions can arise from failure of the immune system to recognize surface-modified cells as autologous, and thereupon attack them as if they were of foreign origin. This constitutes a drug-induced autoimmune reaction. When methyldopa was introduced in the 1960s as an antihypertensive agent, its ability to bind reversibly to the surface of erythrocytes caused this type of haemolytic anaemia to develop in just under 1 per cent of patients receiving the drug over a period of several months. On cessation of therapy, patients made a quick recovery, although antibodies only gradually disappeared. Levodopa and mefenamic acid have caused similar effects.

IMMUNE COMPLEX REACTIONS

These reactions are typified by the serum sickness that frequently occurs during the second week following injection of any serum of animal origin. It is characterized by skin rashes and fever, accompanied by joint swelling and lymphadenopathy. These effects are caused by production of a large, insoluble antigen-antibody complex (I_gG or I_gM immunoglobulins being involved) which damages fine capillaries or causes inflammation. The incidence of serum sickness has been reduced by the introduction of refined antisera which have been treated with proteolytic enzymes such as pepsin to destroy antigenic contaminants. Notwithstanding this, the incidence remains high and animal-derived antisera have been superseded by the use of antibodies of human origin whenever possible. To aid identification of the source, the latter are described as immunoglobulins instead of antisera.

Immune complex formation involving circulating blood platelets can result in thrombocytopaenia after treatment with a variety of drugs, including aspirin, gold salts, phenylbutazone, hydralazine, methyldopa, quinidine, penicillins, sulphonamides and propylthiouracil. It is also responsible for an influenza-like syndrome sometimes seen in patients taking nitrofurantoin, rifampicin, or levamisole, an anthelmintic which is sometimes also used as an immunostimulant in patients whose immune system has been depressed. This may occur after three months' antituberculous therapy in about one fifth of patients who have received more than 20 mg/kg rifampicin once weekly, and is more common in women. It begins an hour or two after a dose is taken, and lasts for about eight hours. Sometimes, it can be stopped merely by changing to a daily dosage regimen.

Systemic lupus erythematosus is an autoimmune disease characterized by deposition of immune complexes in capillaries. Symptoms resembling those caused by it have been observed in about 10 per cent of patients taking large doses of hydralazine over a period of several months. An even higher incidence occurs amongst those who received procainamide for long periods as an antiarrhythmic agent. Anticonvulsants and isoniazid have occasionally produced a similar effect. Antinuclear antibodies have been detected in two thirds of the hydralazine cases, but the mechanism of their formation is unknown.

CELL-MEDIATED REACTIONS

These are typified by contact dermatitis caused by application of any of a wide variety of drugs to the skin. The processes underlying this phenomenon are exceedingly complex, involving transfer of the drug to the surface of T-lymphocytes in the capillaries of sensitized individuals. Within a day or two, increased capillary permeability (possibly caused by release of histamine from mast cells) enables the activated T-lymphocytes to penetrate surrounding tissues and release a group of cytotoxic mediators known as lymphokines. Additionally, the capillary permeability permits infiltration of other cells from the blood, including neutrophils, eosinophils and basophils.

Contact dermatitis has frequently followed sensitization of the skin after repeated topical application of aminoglycoside antibiotics (particularly neomycin, which is now less frequently incorporated in dermatological preparations), sulphonamides, clioquinol, antihistamines and ester-type local anaesthetics such as benzocaine and butacaine. Patients who have become sensitized to these or other drugs should never be given them systemically, otherwise a generalized rash may occur. Some people are sensitized by ingredients such as surfactants and preservatives present in cream or ointment bases.

Some drugs can be converted to reactive intermediates when exposed to light of wavelength in the range 320–450 nm. Formation of such intermediates in the skin will produce photosensitization if they react with tissue protein to activate T-lymphocytes or produce I_gE antibodies. This type of hypersensitivity has been caused by topical application of sulphonamides and ointments containing aminobenzoic acid as preservative, as well as by systemic administration of drugs such as benoxaprofen.

MISCELLANEOUS CASES OF DRUG ALLERGY

It has not always been possible to demonstrate the presence of specific antibodies in patients thought to have become hypersensitive to a drug. This was the case with what was probably one of the earliest drug allergies. In 1923, the popular analgesic-antipyretic, and remedy for gout, cinchophen was shown to be capable of causing hitherto unrecognized fatal liver damage. By that time, it had been on the market for 15 years. During the period 1924–1932 over six hundred million doses were consumed in the United States alone, with 38 deaths attributed to this drug. Since its uricosuric properties afforded relief in gout, and it was safer than colchicine, its use for this purpose was not abandoned until the introduction of safer modern drugs such as allopurinol and the non-steroidal anti-inflammatory agents. Surprisingly, it is still listed in some foreign pharmacopoeias.

The hepatoxicity of cinchophen, which apparently was not dose related, was sometimes associated with skin rashes, suggesting that an immune response was involved. Similar hepatitis-like toxicity has also appeared with the hydrazides used as either antituberculous agents (e.g. isoniazid and pyrazinamide) or as monoamine oxidase inhibitors (e.g. phenelzine), as well as sulphonamides, aminosalicylic acid, ethionamide and phenytoin.

Chlorpromazine and other phenothiazines produce jaundice through intrahepatic cholestasis in a small proportion of patients who become hypersensitized during their first month of therapy. If erythromycin estolate is administered for more than two weeks, there is a possibility that it may do likewise. This does not happen with other erythromycin preparations. Ibufenac, an anti-rheumatic drug, had to be withdrawn by its manufacturer in 1968 after it was found to cause jaundice.

In 1933, a Chicago meeting of the Central Society for Clinical Research was told by F.W. Madison and T.L. Squires that amidopyrine and related compounds might cause agranulocytosis. It was subsequently established that there was a high incidence of this condition amongst doctors and nurses who regularly consumed the drug, whilst another study disclosed that 26 per cent of cases had been preceded by the consumption of it or of closely related drugs. In one group of 128 patients who developed agranulocytosis after exposure to amidopyrine, 70 had died. This led to a clinical trial involving administration of large doses of the drug to over a hundred patients for 40 days or more. When no significant decrease in the number of their leucocytes or granulocytes was found, it was concluded that amidopyrine toxicity was a matter of individual susceptibility. It is now generally believed that hypersensitivity was responsible. In both the United States and the United Kingdom, sales of amidopyrine were restricted by making it a prescription-only drug. Despite this and the fact that the onset of agranulocytosis was sudden and unpredictable, the drug remained popular and appeared in the *British Pharmaceutical Codex* until 1954. Its demise was brought about only by the growing success of one of its analogues, phenylbutazone. As has been seen, this was ultimately shown to cause not only agranulocytosis, but also aplastic anaemia. Sulphonamides, oral hypoglycaemic agents, and antithyroid compounds have also caused agranulocytosis.

Persistent reports of occasional cases of postoperative jaundice in patients who had been anaesthetized with the widely used drug halothane led to the setting up of a National Halothane Study in the United States. The report from this survey of 800 000 patients, which was published in 1966, concluded that halothane was normally no more hepatoxic than other anaesthetics. However, evidence was found of a higher than usual incidence of hepatic necrosis amongst those who were exposed to the drug for a second time within a short period. This has led to speculation that an allergic mechanism is involved. In 1974, Inman and Mushin published their analysis of 130 cases of postoperative jaundice reported to the Committee on Safety of Medicines over an 8-year period. In every case, the anaesthetic had been halothane, and there was strong evidence to support the earlier American findings concerning the dangers of repeated administration. However, the overall incidence of jaundice from this extensively used drug was so low that the only action needed was the avoidance of unnecessarily administering it to patients who had been previously exposed to it within four or five weeks.

Skin reactions may be characteristic for a given type of drug. Common reactions include erythemas, acute urticaria, and pruritis. Although it is unusual for a drug to be withdrawn solely because of adverse effects on the skin, this happened in 1979 with alclofenac, an antirheumatic drug. This was probably because it had no obvious advantage over alternative agents. A United Kingdom trial on 1500 patients had previously revealed that over 10 per cent of them experienced skin rashes. When the UK Product Licence for another anti-inflammatory agent, fenclofenac, was not renewed on its expiry in 1984, this was not simply on account of the frequent reports of skin reactions. The incidence of serious hepatic and renal impairment when compared with the beneficial effects derived from this drug, as opposed to others, was considered by the Committee on Safety of Medicines to be too great. The overall incidence of side-effects had been 14 per cent, albeit the commonest ones being skin rashes of varying severity.

About one in every five patients treated with gold salts experiences skin reactions. These serve as early warnings of hypersensitivity which, if treatment continues, may lead to severe toxicity or even death. Fatal blood dyscrasias can occur.

In 1974, a surgeon at Moorfields Eye Hospital in London, Mr Peter Wright, noticed a markedly increased referral rate to his 'Dry Eye Clinic', the patients presenting with inadequate tear secretion. Extensive inquiries elicited the fact that these patients had taken the antiarrhythmic drug practolol. Shortly after this was reported, dermatologists noted characteristic skin rashes in patients who had been taking practolol. Following the issue to doctors of a warning letter from the manufacturer and the UK Committee on Safety of Medicines, further reports were received of mucocutaneous damage affecting the ears, skin and stomach. Cases of blindness were also reported. The manufacturer responded by restricting supply of practolol to hospital consultants who would prescribe it solely for patients for whom no other therapy was available. When alternative drugs came on the market in 1976, it was withdrawn.

Practolol had been thoroughly tested prior to its marketing. That it took four years before the dangers were recognized was due to its toxic effects being similar to changes which were also seen as part of the ageing process. The incident underlines the fact that no amount of toxicity testing can guarantee the safety of any drug in patients, especially if hypersensitivity is involved. It was not marketed in the United States since it had not satisfied the Food and Drug Administration requirements before its adverse effects were discovered. The mechanism by which it produces its damaging effects remains unknown, but it has been suggested this could involve autoimmunity.

In 1983, the antidepressant drug zimelidine was withdrawn by its manufacturer because ten cases of the Guillaine-Barre syndrome had been reported out of 200 000 prescriptions. This neurological disorder was characterized by progressive muscular weakness. On withdrawal of zimelidine, a slow recovery over a period of months occurred in all cases. Significantly, each patient who developed the syndrome had previously shown signs of hypersensitivity to zimelidine, developing influenza-like symptoms. The incidence of other adverse reactions such as liver damage and convulsions was also relatively high.

VI

Toxicity Testing

When a pharmaceutical company decides to develop a promising chemical compound into a safe and effective medicine, it undertakes an expensive commitment that will be conducted against a background of continuing apprehension lest the project might be abandoned at a moment's notice if unanticipated toxicity problems should arise. No matter how sophisticated drug research becomes, the risk of this will probably never be completely eliminated. Consequently, bringing a new drug to the market inevitably involves something of a gamble. There is no way of telling how many promising compounds fall by the wayside during the early phase of preclinical development, but it must be well in excess of 90 per cent of them. The problem is that the innovative phase of drug research is primarily designed to identify promising compounds whose selectivity of action, or otherwise improved profile, suggests clinical potential. What needs to be determined at the outset is the nature of the toxic effects that might arise when the drug is first tried in humans. It is also necessary to be aware of whether the toxicity profile of a new compound is acceptable for the clinical role it will be expected to fulfil. The processes involved in this have become the subject of strict governmental regulation since the thalidomide disaster.

The thalidomide disaster, apart from being unprecedented in the scale of misery and suffering inflicted, had the most far-reaching consequences of any such event. This was not only because of the severity of the damage, but also because the disaster was experienced in many countries, unlike some of the earlier incidents. Governments throughout the world felt compelled to introduce legislation to control the activity of the pharmaceutical industry as it became apparent that the manufacturers of thalidomide had not carried out the teratogenic testing that was already being undertaken by some leading companies.

Critics of the extensive toxicity testing that has been introduced since the thalidomide disaster are quick to claim that animal tests cannot guarantee the safety of any drug in humans. It has, in fact, never been the intention that these tests should do that! Their purpose is to identify the nature of the toxicity produced by a drug, either when a sufficiently high dose is administered or when seemingly harmless doses are given repeatedly. This not only gives an indication of the probable safe dose range when the drug is administered to humans, but also indicates the potential hazards that might then arise. The critics frequently confuse toxicity testing with safety testing, the latter only being complete after the drug has been fully evaluated on large numbers of patients. Indeed, the safety of a drug cannot be guaranteed even by a clinical trial on hundreds of patients. What happens, in practice, is that as a new drug proceeds through a hierarchy of animal and human investigations, the chances of detecting toxic manifestations increase at each stage. By reducing the number of animal tests below a critical level, the risks faced by the first human volunteers and the patients in the early trials are thereby increased. How to strike the right balance between too much and too little animal testing must remain a matter of fine judgement. The tendency is to err in favour of excessive testing through fear of litigation in the event of a subsequent mishap.

Animal tests can mislead not only by failing to detect toxicity, but also by wrongly implying that a drug is unduly toxic. Although this is more likely to happen with carcinogenicity testing, it applies to other areas as well. It was fortunate that Florey and Chain did not decide to use guinea-pigs when first testing penicillin, for they may then have abandoned the project as these animals are hypersensitive to penicillin. Herein lies the fundamental problem in toxicity testing. What significance is to be attached to adverse reactions in animals?

ACUTE TOXICITY TESTING

The first toxicological studies on any new compound involve the administration of single doses to rats, followed by a repeat of the experiment at increasing dose levels until the maximum tolerable dose is ascertained. The animals are kept under observation for at least 14 days to allow for delayed toxicity. This type of limit testing can provide an early indication of the safety margin between the projected therapeutic dose and that which will produce toxic effects. It can also reveal what might happen if a patient received an overdose. The signs of toxicity which the trained observer will notice include those caused by action of the drug on various parts of the central nervous system. These include dyspnoea, ataxia, emesis, convulsions, tremor, lethargy, sleep, coma and behavioural changes. Toxic effects on the autonomic nervous system are revealed by cardiovascular changes, alteration in muscle tone, gastro-intestinal disturbances such as diarrhoea, excessive salivation, piloerection, and also by ocular effects such as lacrimation, ptosis, relaxation of the nictitating membrane in the cat, and miosis or mydriasis. Toxicity to particular organs may be detected by superficial observation, as when kidney injury leads to production of reddened urine or oedema. More frequently, organ damage is recognized on histological examination of the tissues after death of the animal.

It has long been the purpose of acute testing to indicate whether a new compound is worthy of further consideration as a candidate for development. In practice, most new compounds fail to pass this hurdle. It has been estimated that as many as ten thousand compounds are synthesized for each drug that ever reaches the market! Since the introduction of modern regulatory controls, acute toxicity testing has been required by drug regulatory authorities before they will grant permission for a clinical trial to begin. The detailed regulations vary somewhat from country to country, often necessitating the repetition of the tests if a new submission is made to a different authority. Those who have the welfare of animals at heart would be well advised to pressurize their governments to seek international agreements ending this unnecessary state of affairs. In the United Kingdom the regulations require a new drug to be studied in at least two mammalian species. In North America three species are exposed to three dose levels known to produce varying degrees of toxicity. This is designed to establish in detail how the pattern of toxicity varies with dose. One of the species must be a non-rodent, dogs or monkeys often being preferred. The animal most frequently used in acute toxicity testing is the rat. Allowance has to be made for the fact that in rodents the way a drug is metabolically detoxified can be influenced by the sex of the animal. Thus toxicity studies are conducted on both male and female rodents. In addition, the effect of the new drug is studied in both very young and old rats in order to obtain some indication of problems that might arise when the drug is administered to children or elderly patients.

Much criticism has been levelled at the use of lethality testing of new drugs as this affords little information about the toxicity of sub-lethal doses. Lethality testing was introduced in 1927 by Trevan, who used frogs to compare the lethal dose of powdered digitalis samples with that of a standardized preparation. This permitted the potency of different batches of the powder to be assayed with considerable reliability. Over the years, however, the conception of this type of test has changed from comparative to absolute insofar as it has become widely accepted as a criterion for judging drug toxicity. This involves the LD_{50} test, which provides an estimate of the lethal dose which kills 50 per cent of a group of at least six animals. Since lethality varies for different routes of drug administration, it is mandatory for the LD_{50} value to be calculated for all those proposed for human therapy. Regulatory authorities accept many of the criticisms of lethality testing and thus do not insist that the LD_{50} value be precisely determined. It is sufficient for it to be calculated by extrapolation from the figures compiled during limit testing.

REPEATED DOSE TOXICITY TESTING

Most drugs are taken repeatedly, hence more extensive toxicity testing than that just described is essential. Much valuable information can be gleaned by examining the effects of repeated doses of a test compound on animals to find which organs are damaged by it. As with acute toxicity testing, the intention is not to prove that a drug is absolutely safe, but rather to identify the nature of the toxic reactions that might occur in humans. The regulatory requirements in most countries call for the tests to be carried out in two species, one of which must be non-rodent. The choice of species is influenced by any information that is available indicating how the drug is absorbed, distributed, metabolized and excreted in human volunteers. The objective is to select a species of animal which will absorb and dispose of the drug in a similar manner to that pertaining in humans. The plasma half-life (i.e. time taken for the plasma drug level to be halved) is a valuable parameter to determine since it often reflects the pattern of absorption, distribution, metabolism and excretion of the drug. Unfortunately, this parameter can vary greatly within groups of both humans and animals treated with identical amounts of the drug in question. Certain species also present specific problems. For example, rats will excrete an acid in the bile if its molecular weight is over 300, whereas in humans this usually happens only for acids with a molecular weight in excess of 500. For a variety of reasons, rats and dogs often prove to be suitable for repeated dose toxicity studies. Monkeys are also used.

The duration of repeated dose toxicity testing reflects that of the intended clinical usage of the drug. Regulatory authorities issue guidelines that assist manufacturers in determining this. For example, a drug that is to be administered for not longer than one week will be given daily to animals for 28 or 30 days whereas one that is to be taken for up to a month must be tested for at least 90 days. Naturally, the route of administration during toxicity testing must match that intended for therapeutic purposes; tests must be repeated for different routes. If a drug is going to be taken by mouth, it can be given by gavage (forced feeding) or it can be added to the diet of test animals for a major part of their life span. However, when a drug is to be injected parenterally, regulatory authorities use their discretion to make allowance for the inevitable irritation that would be caused by repeated injection of relatively large volumes into rodents.

Repeated dose toxicity testing, which is also described as sub-acute toxicity testing, is not only conducted at dose levels which produce something akin to the desired clinical response, but is also carried out at dose levels which induce definite signs of toxicity such that a small proportion of the animals might even die before completion of the tests. Testing is also carried out at intermediate dose levels. As with acute toxicity testing, its value is determined by the vigilance exercised by the investigators. Guidelines are again laid down by regulatory authorities, those in the United States being particularly exacting as a consequence of legislative action (*viz.* Good Laboratory Practices) introduced in 1979 in the wake of a scandal involving alleged faulty procedures by one pharmaceutical company. Thorough histo-pathological examination is required on all organs of the animals sacrificed at the end of the test period. This means that literally thousands of microscope slides have to be prepared and examined. It is also essential to observe the living animals closely and monitor as many body systems as is practicable. This is why the use of large animals such as dogs is preferred. Particular attention is paid to body weight, a sensitive indicator of animal health. Routine checks on food consumption and general behaviour are supplemented by detailed analysis of blood and urine, together with specific tests predicated by the nature of the drug being studied. Particular attention is given to measurement of renal clearance rate, and electrolyte levels. Possible hepatic toxicity is checked by monitoring the enzymes released into the blood when liver cells are damaged. The precise nature of the damage can often be determined by differential examination of a range of these

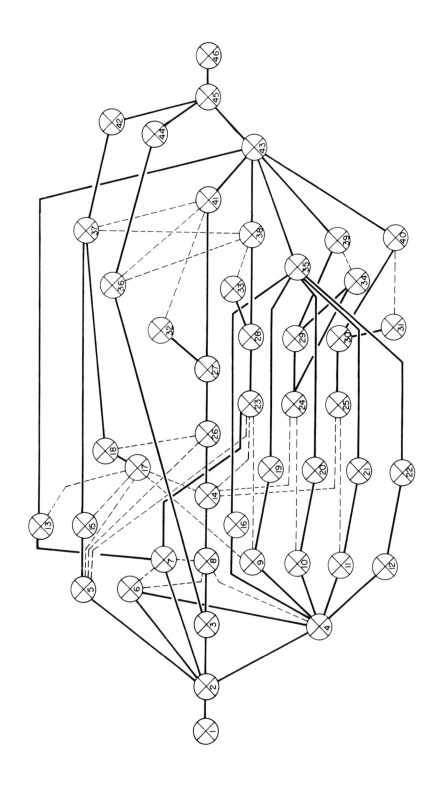

Nodes

(1) Decision to take special interest in compound active in screening test
(2) Decision to involve other departments
(3) Completion of main pharmacological work
(4) End of preliminary toxicity
(5) Completion of assay mtheod
(6) Decision re dose, route etc, by medical department
(7) Decision re dose form by pharmacy
(8) Beginning of preliminary subacute rat test
(9) End of acute tests
(10) End of acute rabbit test
(12) End of rest of acute tests
(13) Completion of irritancy test
(14) End of preliminary subacute rat test
(15) End of preliminary drug metabolism studies
(16) End of histology
(19) End of histology
(20) End of histology
(21) End of histology
(22) End of histology
(23) Beginning of subacute dog test
(24) Beginning of teratogenicity tests
(25) Beginning of another subacute test
(26) Beginning of main subacute rat test
(27) End of subacute rat test
(28) End of subacute dog test
(29) End of teratogenicity tests
(30) End of another subacute test
(31) Completion of histology
(32) Completion of histology
(33) Completion of histology
(34) Completion of skeletal examination
(35) Completion of acute toxicity tests
(36) End of pharmacology
(37) End of drug metabolism studies
(38) Completion of subacute dog test
(39) Completion of teratology
(41) Completion of subacute rat test
(42) Completion of drug metabolism submission
(43) Completion of toxicology submission
(44) Completion of pharmacology submission
(45) Complete report assembled
(46) Report passed to registration department

Activities

1– 2 Extra pharmacology (construct logic diagram?)
2– 3 More pharmacological tests
2– 4 Perform preliminary acute toxicity tests
2– 5 Perfect method for assay of compound in biological fluids
2– 6 Consult medical unit
2– 7 Consult pharmacy unit
3–36 Finish and write up pharmacology
4– 6 Transmit information to medical department
4– 9 Perform acute dog test
4–10 Perform acute rabbit test
4–12 Perform rest of acute tests
4–16 Perform histological examination
5–15 Perform preliminary drug metabolism studies
7–13 Perform local irritancy test
7–23 Pharmacy prepare material for dogs
8–14 Perform preliminary subacute rat test
9–19 Perform histology
10–20 Perform histology
12–22 Perform histology
13–43 Write up irritancy test
14–26 Wait and plan
15–37 Complete drug metabolism studies
16–35 Write up preliminary acute tests
19–35 Write up actue dog test
20–35 Write up acute rabbit test
21–35 Write up another acute test
22–35 Write up rest of acute tests
23–28 Perform subacute dog tests
24–29 Perform teratogenicity tests
24–34 Begin skeletal examinations
25–30 Perform another subacute test
26–27 Perform subacute rat test
27–32 Perform histology
27–41 Compute and write up
28–33 Perform histology
28–38 Compute and write up
29–34 Finish skeletal examinations
29–39 Compute and write up
30–31 Perform histology
30–40 Compute and write up
35–43 Compile for toxicology submission
36–44 Write pharmacology submission
37–42 Write drug metabolism submission
38–43 Compile for toxicity submission
39–43 Compile for toxicity submission
41–43 Compile for toxicity submission
45–46 Perform last revision

Notes: In addition, the following should be added in more detailed critical path analysis:
(i) Before (2), information on basic physical and chemical characteristics of material, and on stability.
(ii) Before (8), information on purity and nature of impurities.
(iii) Identify quantities of drug substance/product required for activities 1–2, 2–3, 2–4, 2–5, 3–36, 4–9, 4–10, 4–11, 4–12, 5–15, 7–13, 7–23, 8–14, 15–37, 23–28, 24–29, 25–30, 26–27.

Figure 8. A network analysis detail for pharmacological and toxicological clearance of a new drug substance. *Reproduced by permission of The Pharmaceutical Journal and K.A. Lees.*

enzymes appearing in the serum. A variety of methods also exist for detecting gastro-intestinal bleeding through measurement of faecal occult blood. The assessment of toxicity to the eyes, lungs, skin and other tissues also involves specific test procedures. Microscopic examination of the spleen, thymus and lymph nodes can be used in an attempt to make some assessment of the action of drugs on the immune system.

Broadly speaking, the repeated dose testing procedures are similar throughout the world, but minor variations present problems for pharmaceutical manufacturers seeking approval for their products in different countries. In addition to the toxicity tests already described, there must be studies of the action of any new drug on fertility and on the foetus. The latter studies are extended to include careful monitoring of the new-born animals until after the period of lactation, followed by pathological investigation.

TERATOGENICITY TESTING

Prior to the thalidomide disaster, teratogenic studies were carried out only on drugs that were likely to be taken during pregnancy or which had hormonal effects. The tests were based on foetal survival, the assumption being made that any drug-damaged foetus would be aborted. Inadequate as this turned out to be, when thalidomide was later subjected to this type of testing it was found to reduce litter size in female rodents exposed to large doses of drug before mating and during pregnancy. Although there was no evidence of malformation in the surviving new-born rats, the thalidomide disaster might still have been avoided if those manufacturers who advertised it for use in pregnancy had insisted on this type of test being carried out. Reduced litter size would surely have made them think again about promoting the drug in this manner.

Strict requirements now exist for teratology studies, these specifying the number of species, periods of exposure to the drug, the stage at which the foetus is removed, and the like. Careful consideration is given to the dose level so that it accords with that likely to be used in humans. Favoured animals are the rat and the rabbit. Basically, the tests fall into three phases. The first is designed to assess effects on fertility and general reproductive performance in test and control animals. This involves observing drug effects on the gonads of both sexes, the oestrus cycle, mating frequency and conception rates. Typically, males are administered the drug for 60 days, females for 14 days, then mating is permitted. After this, the males are killed and their reproductive organs are examined. The females continue to receive the drug throughout pregnancy, half of them being killed and examined at the mid-point of the gestation period (i.e. the eleventh day in rats). The remaining females are administered the drug until they are sacrificed and examined at the beginning of the weaning period. The second phase of the study is designed to detect teratogenicity. In this, pregnant animals are dosed throughout the period of organogenesis, then on the day before parturition foetuses are removed by section and examined for signs of skeletal or visceral abnormalities. When assessing the significance of any abnormality that is found, due consideration is given to the background incidence of it in unexposed foetuses. The third phase of investigation deals with perinatal and postnatal toxicity. This time, the pregnant animals receive the drug in the final third of the gestation period and throughout the period of lactation. The young animals are then killed and thoroughly examined.

CHRONIC TOXICITY AND CARCINOGENICITY TESTING

Once approval to begin tests of a candidate compound on patients has been granted by a regulatory authority, it becomes essential to initiate tests that will establish the long-term safety of the drug. As has been indicated, the purpose of acute and repeated dose toxicity testing is to establish the nature of the toxic effects that will inevitably occur if sufficiently large doses of any drug are given. The possibility of harmful effects arising from repetitive administration of a drug at apparently safe dose levels cannot be excluded by such tests. Hence it is necessary to conduct long-term studies in animals using doses that compare with those considered likely to be safe when given to patients. Since it is difficult to be certain what single dose level should be employed, regulatory authorities once again call for the tests to be repeated at three or four different levels. The highest level should produce the first signs of reversible toxicity in some animals. The monitoring tests previously described are carried out at frequent intervals, and periodically a few animals are sacrificed for detailed pathological investigation. The detailed studies are identical to those for repeated dose toxicity testing.

All regulatory authorities are understandably concerned about the risk of a new drug proving to be carcinogenic. Any compound whose molecular structure, in whole or in part, in any way resembles that of a known carcinogen is immediately suspect and must be thoroughly evaluated. Since it is generally recognized that the hazard from carcinogenicity increases with the frequency of intended use of the drug, most authorities call for tests on any drug which might be administered to patients for more than six months. Furthermore, if the toxicity testing or the nature of the biological action of the drug suggests it has any action on nucleic acids or the process of cell division, carcinogenicity testing becomes essential. The tests are done on rodents and follow the general pattern for chronic toxicity testing. When selecting two animal species, care is taken to avoid those with a high or variable incidence of spontaneous tumours. Mice, rats and hamsters are widely employed, but these might not really be suitable predictors of what could happen in humans. The detection of breast cancer from exposure to high-potency progestogens was originally made in beagle dogs, not rodents. Whatever species is finally selected, the problem of what significance to attach to the results is perplexing. How great an increase in the natural incidence of spontaneous tumours can be considered statistically significant, and what allowance should be made for the facility with which some inbred strains of animals develop tumours? What strains are appropriate—

those which are more or those which are less resistant to tumour induction?

Care must be taken in selection of an appropriate dose that is high enough to increase the likelihood of inducing cancer, yet does not produce sufficient toxicity to kill the animals before the testing is completed. Exposure of the animals to the test compound is initiated at weaning and continues for most of their life span (i.e. 18 to 24 months in mice, and sometimes even longer in rats). If there is significant prolongation of life when exposure is stopped earlier, some animals can be treated this way in order to assess whether delayed appearance of a tumour occurs. In general, five hundred rodents may be required to ensure survival of enough members from both exposed and control groups, hence the cost and time that is involved is considerable. Pharmaceutical companies cannot afford to wait for the completion of these tests before forging ahead with their developmental and early clinical studies.

For many years it has been believed that the development of cancer may be preceded by mutagenesis. In 1973, B.N. Ames and his colleagues showed that 85–95 per cent of carcinogenic substances were capable of producing readily detectable genetic mutations in a test which exploited the inability of a mutant strain *Salmonella typhimurium* to synthesize histidine. This aminoacid was essential for growth and reproduction. If the bacteria became able to grow and multiply in a histidine-free medium after exposure to a test drug, this was probably due to an induced genetic mutation. There has been considerable interest in mutagenicity testing as an alternative to expensive, time-consuming carcinogenicity testing. Such tests have been carried out at an early stage in the development of new drugs in an attempt by manufacturers to avoid problems arising later when the results of conventional carcinogenicity tests become available. Recent developments have meant that up to two hundred substances can now be tested in one day, growth of the bacteria being detected turbidimetrically by an optical device.

The problem with mutagenicity testing is that not only are some carcinogens not mutagenic, but there are also mutagens which are not carcinogenic. Nevertheless, since the autumn of 1985 mutagenicity testing has been mandatory in member countries of the EEC wishing to issue licences for new medicinal preparations. This is a welcome move because the real significance of mutagenicity testing lies in its ability to forewarn of the potential of some chemicals to induce heritable mutations.

Clinical Trials

Whether the extensive toxicity testing of new drugs in animals is altogether justifiable in terms of the information gained is debatable. There is a feeling in some quarters that the political response to the thalidomide tragedy has forced government agencies to become over-zealous in their adoption of many current requirements. The media criticism of the Committee on Safety of Medicines for its handling of the benoxaprofen incident may have been ill-informed, but it will certainly have made that body even more cautious than before. The chances are that patients will be left to die while pharmaceutical companies are forced to delay the introduction of life-saving drugs so as to satisfy demands for a level of safety that is not expected from therapeutic measures such as surgery, or from products of the automobile and aircraft industries. The moral stand of a society which seeks to prevent the exposure of its sick citizens to the low level of risk from new pharmaceutical products, whilst simultaneously turning a blind eye to the sale and advertising of the most deadly drug ever marketed is, to say the least, perplexing! Tobacco products kill one in every four of their regular users.

Many responsible scientists within the pharmaceutical industry and in academia believe that early studies in humans are of more value than extended tests on animals whose metabolism is quite different from that of *Homo sapiens*. In practice, what happens today is that as toxicity testing in animals progresses through its various phases, sufficient data emerge to permit experiments to be safely conducted on healthy human volunteers (no attempt is ever made to expose them to cytotoxic drugs intended for use in cancer patients). The first of these short-duration experiments can begin after completion of the acute toxicity studies. This earliest trial of a new drug on humans involves administering to twenty or so volunteers a single dose or, if sufficient animal data have already been acquired, a small number of repeated doses. In some countries, permission must first be obtained from the regulatory authority. In America, for example, manufacturers have to submit to the Food and Drug Administration a Notice of Claimed Investigational Exemption for a New Drug (IND) application before proceeding to test drugs on volunteers.

Most of the volunteers who participate in drug trials are employees of pharmaceutical companies. In the United Kingdom, the occasional paid employment of students for such work has been the subject of some criticism, especially as no statutory controls exist, in contrast to the situation in the United States. Nevertheless, volunteers are subjected to a thorough medical examination before tests begin, and are kept under strict surveillance at all times. To ensure that specialist medical attention is immediately available in

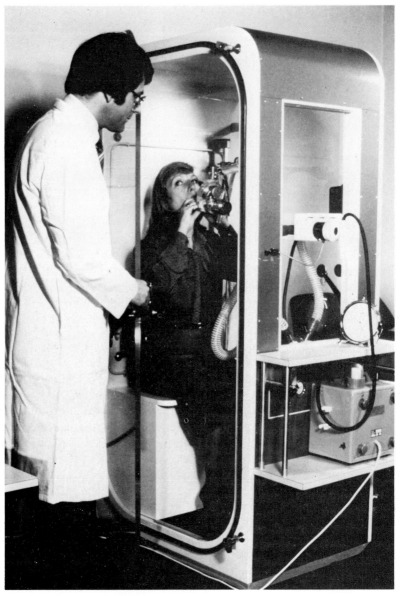

Figure 9. Measuring the effect on lung function of an inhaled drug on a healthy volunteer. *Reproduced by permission of the Glaxo Group of Companies.*

the event of any untoward reaction, the trial should be conducted in a hospital or similar environment.

The objective of the volunteer studies is to obtain an idea of how the human body deals with the new drug which, after all, is a foreign compound. The first dose given to any volunteer is always a fraction of that previously found to be safe when administered by the same route to animals; this would usually mean something in the order of from one tenth to one fiftieth of that of the animal dose once allowance has been made for the differences in body weight. If all goes well, the dose can be cautiously increased on subsequent occasions. This will furnish an indication of what dose can eventually be given to the first patients who receive the drug. Naturally, the amounts of the drug appearing in the plasma and urine of the volunteers are carefully monitored, either by a chemical assay or by utilizing a radio-labelled form of the drug. Repeated blood counts and monitoring of both hepatic and renal function are essential in view of past experience of likely adverse drug reactions. At the end of the day, if the absorption, disposition and fate of the drug being investigated turn out to be similar to that previously observed in animals, greater reliance can be placed on the predictive value of the animal results.

Comparison between the blood levels attained in volunteers after oral and intravenous dosing is important. If adequate blood levels of a drug were only achieved by the latter route, it would suggest that the fault lay in ineffective absorption from the gut rather than rapid metabolic inactivation. Alternatively, if reasonable plasma levels were observed after oral administration, it would be possible to assess the degree of bioavailability by that route (the detailed mathematics of such assessments lie within the province of pharmacokinetics and need not be dwelt upon here). Examination of the plasma and urinary levels of the compound and any of its metabolites which can be detected provides valuable information regarding their fate.

Drugs such as antibiotics, antirheumatic agents or antidepressants may not produce any detectable effects in healthy individuals. However, those which do produce a pharmacological response in healthy volunteers may provide valuable information relating to their clinical potential. A pharmacological effect can sometimes be related to increasing plasma levels of an active metabolite which is more potent than the parent compound. Such a finding may be exploited through the manufacture and testing of this metabolite. Conversely, the display of adverse effects associated with the appearance of a toxic metabolite may require the synthesis and screening of analogues that cannot form it.

Tests of a new drug on healthy volunteers may sometimes suggest that it is, or is not, likely to be of clinical value, but there always remains the possibility that either the action, absorption, disposition, metabolism or elimination of the drug may be modified by the disease process. Some antibiotics, for example, will penetrate into the cerebro-spinal fluid more readily than otherwise if the meninges are inflamed. It is prudent to initiate preliminary studies in patients as soon as the results from chronic toxicity testing and volunteer studies indicate that there are no serious hazards likely to be encountered. In the post-thalidomide era it has become necessary to receive formal permission from the regulatory authority before these preliminary studies can be initiated. As with other aspects of drug licensing, the requirements vary from country to country, but tend to follow the American lead. In the United States, preliminary investigations in patients are covered by the Investigational New Drug application. For this, full details of all toxicity testing and product quality assurance results will have been submitted. Much of the responsibility for giving approval for clinical trials is actually devolved to Institutional Review Boards, the constitution of which are regulated by the Food and Drug Act. The Food and Drug Administration retains the right to prevent a trial proceeding, so long as it notifies the applicant within 30 days.

Before clearance to administer a new drug to patients in the United Kingdom can be granted, a pharmaceutical company must supply to the Medicines Division information similar to that required by the Food and Drug Administration in the United States. The professional staff in the Medicines Division must then satisfy themselves that the formulation of the drug is consistent and of a high quality, that the chemical assay procedures are reliable, and that no particular hazard has been identified during toxicity testing. This being so, a Clinical Trial Exemption Certificate will be granted. If there is any uncertainty, the application is refused. The manufacturer then has the option of formally applying to the Committee on Safety of Medicines for a Clinical Trial Certificate. This body may decide that further information must be provided before the certificate can be issued. Permitting the full-time staff of the Medicines Division to issue exemption certificates has the merit of avoiding unnecessary delays in the initiation of clinical trials since a decision has to be reached within 35 days. This flexibility is made possible by the fact that almost all clinical trials have first to be approved by an Ethical Committee. In contrast to American practice, the constitution of such committees varies considerably; the British Medical Association recommends that there should be eight members, namely two senior hospital doctors, one junior hospital doctor, two general practitioners, a community medicine specialist, a nurse and a layman.

The decision to test any new drug on patients presents an ethical problem since neither its absolute safety nor its efficacy can ever be guaranteed. If beneficial results can be obtained with any existing form of therapy, what right does the physician have to abandon this in order to test a new drug of no proven merit? Alternatively, at what point is existing therapy considered to be of such dubious value that a promising new drug merits a trial in a patient who has placed his trust in the physician? This question must surely be highly pertinent in view of the fact that many new drugs offer no significant advantage over existing agents. It could, of course, be argued that the physician has to consider the overall benefit to suffering humanity at large, rather than merely consider the needs of an individual patient. But would that not contradict the Hippocratic Oath?

Nor can the legalistic formulation of 'informed consent from the patient' really enable the physician to avoid the moral dilemma; few patients are truly informed about the issues that are involved. The situation is even more complex in children, the mentally retarded and the elderly. Little wonder, then, that patients who fail to respond to existing therapy are usually the first to be entered in a clinical trial. In the final reckoning, society must place its trust in the humanity, wisdom and judgement of those responsible for the care of the sick.

The earliest clinical studies on a new drug are primarily devised to elucidate the pharmacological actions of the drug and reveal whether its absorption, distribution, metabolism and excretion follow the pattern previously observed in healthy individuals. Such studies may involve from 25 to 50 patients afflicted with a condition which is likely to be alleviated by the proposed therapy. Any evidence relating to the clinical efficacy of the drug would be considered a bonus at this stage. Much importance is also attached to the outcome of pharmacokinetic studies as these can reveal whether there is likely to be toxicity from accumulation of the drug if a given dose is administered too frequently. What usually happens is that after a few patients have received small doses without untoward reactions, sufficient confidence is gained and more patients are then given larger doses. However, the question of appropriate therapeutic dosing cannot be finally settled at this stage of the investigation.

In the United States, the Food and Drug Administration describes the combined investigations in healthy volunteers and the preliminary clinical pharmacology in selected patients as a Phase I study. Whilst this term has gained general acceptance, confusion has arisen in the United Kingdom as to whether it includes the first studies on healthy volunteers since the Medicines Act only requires a licence application to cover investigations on patients. There is no such ambiguity over what is meant by either a Phase II or a Phase III study. The former refers to the early assessment of drug efficacy and appropriate dosage for sick patients, whilst the latter covers the extensive clinical trials in hundreds of patients, this being necessary before regulatory authorities will issue a Product Licence permitting the drug to be marketed. In the United States, clinical trials conducted after a drug has been put on the market are described as Phase IV studies.

The number of patients involved in the Phase II studies varies widely, depending on the nature of the medical condition being treated. As a rough indication, perhaps 150–350 people may receive the drug. The investigation aims to determine a safe dose schedule and how long it will be continued in the full-scale clinical trials, as well as to identify which type of patient responds adequately to the drug. Every effort is made to identify the inevitable side-effects, due regard being given to the indications from the chronic toxicity studies in animals. From these studies, it is hoped to optimize the administration of the drug so that it will be used to full advantage in carefully controlled clinical trials. Pharmaceutical companies meet the cost of such time-consuming investigations, recognizing their immense value in ensuring either the withdrawal of inferior products before an expensive commitment to proceed is made, or—and this is less frequently the case—in obtaining the maximum value from the candidate drug.

The decision to proceed into Phase III can only be taken if there is clear evidence that a drug is effective at a tolerable dose level. There is immense institutional pressure on the management of pharmaceutical companies to take this step even when there is no proof that the drug is superior to rival products already on the market. It is often the case that any comparison with other medicines will only have been based on the subjective views of the clinicians; these can be dismissed or accepted with equanimity. Only a statistically controlled large-scale clinical trial comparing the new drug with the best available rival product can settle the issue. This is why so many drugs of no proven superiority reach the market. Few regulatory authorities are free to demand that a new product must have proven superiority; only safety and efficacy may be considered. At first sight, this attitude appears to ignore the fact that the safety of a new product cannot be properly assessed until it has been used routinely by possibly hundreds of thousands of patients. Is it justifiable to expose the public to the slight risk of unanticipated toxicity from a product with no proven advantage over existing agents? Further reflection on this matter suggests that the regulatory authorities wisely recognize the limited value of comparative clinical trials when assessing efficacy. Firstly, the numbers of patients involved are restricted, seldom being more than one or two thousand, and often much less. Larger trials may have to be conducted on a multicentre basis, creating the problem of variation in interpretation of the clinical results at the different centres. Furthermore, there can be difficulty in organizing a close match of all the factors that might conceivably influence the outcome of the treatment. For example, the age distribution of the patients may not reflect that of the sick population likely to receive the drug (does society accept the wisdom of testing new antiarthritic drugs in octogenarians?). Then there is the question of the duration of clinical trials. Would apparently comparable drugs appear so if they could be tested for the ten years or longer for which antihypertensive and antiarthritic agents are used by many patients? Finally, even the parameters selected to monitor the efficacy of the drug may not reflect all the important criteria for different patients. One can visualize the spectrum of views held on the merits or otherwise of different psychopharmacological agents ranging from mild tranquillizers to monoamine oxidase inhibitors used as antidepressants.

The aims of a clinical trial must be clearly defined at the outset. Considerable care is then required in order to ensure that the results of the proposed trial will be statistically valid. The expert advice of a statistician is always essential at this stage, for no amount of mathematical manipulation can rectify a badly designed trial. Close collaboration between clinicians and the statistician can help to keep the number of patients exposed to the potential hazards of an unproven therapy to the minimum necessary for statistical validity of the particular type of measurement of efficacy that is selected. The chosen parameter for assessment is not always amenable to precise measurement, and even if it is, there is usually fluctuation even in the absence of treatment. Thus it is essential to establish whether the apparent benefits conferred by treatment are genuine or merely due to chance. This is indicated in clinical reports by the use of the probability coefficient p. A value of 0.05 indicates there is a 5 per cent probability of the observed beneficial action of a new drug in fact being due to chance. Obviously, there can be much greater confidence when the value is 0.01.

Clinical trials often compare a new drug with an existing agent. The choice of a standard for comparison is crucial, for a disreputable manufacturer could gain much by advertising the superiority of a new product over a substance that those without adequate expertise might mistakenly believe was the drug of choice. A valid comparative trial not only has to avoid intentional bias, but also that which is unintentional. Any indication that tells either the physician or the patient which drug is being administered must be avoided if at all possible. There is no shortage of evidence from past trials to show that patients are highly susceptible to suggestion. The necessity of presenting both drugs in identical dosage forms is obvious; this should extend to cover taste and any other relevant feature. The patient must remain blind as to the nature of the medication, but so too must the clinical attendants. This is achieved by the 'double blind' technique, whereby a pharmacist ensures that neither medical nor nursing staff have any idea which of the drugs is being administered, the labelling code being sophisticated enough to hide this information. Ideally, all patient records should be maintained and kept up to date by the pharmacist. Bias in selection of patients for each assessment group is avoided by strict adherence to a statistically valid randomization procedure. In clinical trials where the underlying disease process remains unaltered by the treatment and will reappear soon after withdrawal of therapy, it is possible to apply a crossover technique in which the treatment schedules are periodically reversed. There are, however, only a limited number of situations where this can be reliably done. It should go without saying that whatever technique is chosen when comparing a new drug with an established one, professional integrity demands that every effort be made to use the established drug in the most efficacious manner that can be achieved. Sub-optimal dosing with regard to amount, frequency or duration will invalidate the result of any clinical trial.

Where a drug has a unique therapeutic activity, a comparison has to be made against an inert placebo which has been carefully formulated to appear identical. A placebo may be included in trials where an established drug is being used as a standard for comparison, so long as patients suffer no lasting harm by this denial of therapy. This may be apposite in those instances in which the established treatment itself is of dubious value. An ethical problem arises if patients with a life-threatening disease are to be assigned to receive a placebo. Can a patient with a tumour that has, say, only a 10 per cent chance of being cured with a combination of drugs be given a placebo in order to evaluate a new cytotoxic compound? Surely a 10 per cent chance of survival is better than none at all? Clinical oncologists try to avoid this type of problem by instead studying the effect of adding a new drug to an existing combination of proven efficacy.

VIII

Regulatory Controls over New Medicines

Before a new pharmaceutical product can be introduced into any of the leading markets it must first receive approval from the appropriate national drug regulatory authority. Whilst the procedures vary from country to country, there are broad similarities in the processes that must be followed. In general, regulatory authorities are concerned with the quality, safety and efficacy of medicines for human consumption.

In 1875, the government of Benjamin Disraeli introduced its Sale of Food and Drugs Act which resulted in the establishment of an inspectorate and analytical laboratories. This effectively curtailed any attempts at the deliberate adulteration of drugs or the sale of medicines containing ingredients known to be harmful. The Act was responsible for encouraging the scientific training of professional analysts. It also stimulated the US Department of Agriculture to seek similar legislation to control the outrageous behaviour of itinerant salesmen of an estimated 50 000 patent medicines which at best were quite ineffective and at worst were lethal. Yet it was not until 1906 that Congress passed a Food and Drugs Act to put an end to the adulteration of drugs involved in interstate commerce. The limitations of this Act became obvious when it failed to prevent the disaster in which the diethylene glycol in Elixir of Sulphanilamide killed 107 Americans in 1937. The following year, the Food Drug and Cosmetic Act was modified so that it not only controlled the registration and inspection of the premises of pharmaceutical manufacturers, but also required clinical evidence of safety when a medicinal preparation was used in accordance with the directions on the label. Manufacturers now had to submit a New Drug Application (NDA) before a product could be marketed. Twenty years later, it was this legislation which prevented thalidomide from being sold in the United States before its teratogenicity was recognized. Other countries paid a high price for not having followed the American example. They were finally compelled to react when faced with the grim reality of the thalidomide disaster.

There was an immediate reaction to thalidomide in the United States. In 1962, the Kefauver-Harris amendments to the Food, Drug and Cosmetic Act added new provisions which had far-reaching consequences. These were largely designed to ensure much more extensive preclinical evaluation of drugs (chemistry, pharmacology, toxicology, formulation data, quality assurance and manufacturing procedures) and more thorough clinical investigations. To this end, the latter had to be registered through the submission of a Notice of Claimed Investigational Exemption for a New Drug (IND) which clearly delineated the purpose of the proposed investigation. In addition, informed consent had to be obtained from volunteers and patients participating in clinical trials, whilst an assurance of their thorough supervision was necessary. A major innovation that, for the first time, manufacturers had to submit convincing proof of efficacy in addition to safety when proceeding with an NDA submission. Further controls over the marketing of new drugs were also instituted. Various committees had to be established to deal with the differing aspects of this extra work generated by the Kefauver-Harris amendments. The Office of Drugs was given the overall responsibility for matters concerning medicinal products, whilst the Office of New Drug Evaluation dealt with the INDs which had to be approved before clinical studies on a new drug or formulation could begin. It also handled the NDAs that followed if the results of these trials convinced the manufacturer that the drug should be marketed. Once an NDA was approved, the manufacturer was free to sell the drug in the United States, but regular reports had thereafter to be submitted to the Food and Drug Administration. The Office of Biologics carried out similar duties for biological products such as vaccines, sera and blood products. In 1966, the Food and Drug Administration extended the scope of its requirements to cover non-prescription drugs as well, and it initiated an ongoing review of over three thousand existing drugs which had not been evaluated for efficacy.

Many countries, especially those whose own manufacturers wish to sell pharmaceuticals in the United States, have based their drug regulatory legislation on the American model. This entails the entire gamut of acute and chronic toxicity testing, quality assurance, clinical trials, good manufacturing practice, factory inspections and all else that is now deemed necessary before a medicine receives approval for marketing.

The thalidomide episode exposed the weakness in the British system of leaving the control over the introduction of new drugs to the good sense of the pharmaceutical industry. Nevertheless, there was still a considerable element of trust remaining between government and industry, so the initial reaction to this bitter experience was to establish the Committee on Safety of Drugs, a government-appointed agency that secured the voluntary cooperation of more than six hundred companies who agreed to accept its advice and decisions prior to starting clinical trials or marketing any new product. The committee adopted most of the procedures being followed in the United States, although it did not insist on approval being sought for the early studies on healthy volunteers. Regrettably, a number of small manufacturers refused to cooperate with the Dunlop Committee, as it became known—the chairman being Sir Derrick Dunlop. This might have been ignored had it not been for mounting public concern about the responsibilities of the pharmaceutical industry, in part provoked by the protracted legal proceedings concerning the question of compensation for thalidomide victims. In 1968, the government introduced the Medicines Act, which amongst other measures established the Medicines Commission and replaced the Dunlop Committee by a statutory Committee on Safety of Medicines (consisting of all the members of the earlier committee).

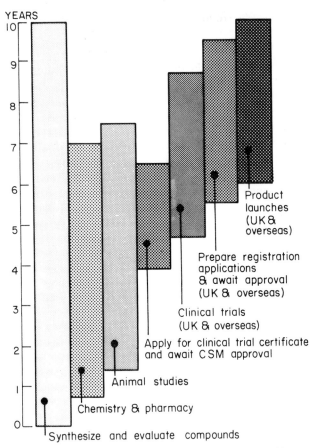

YEARS

Product
launches
(UK &
overseas)

Prepare registration
applications
& await approval
(UK & overseas)

Clinical trials
(UK & overseas)

Apply for clinical trial certificate
and await CSM approval

Animal studies

Chemistry & pharmacy

Synthesize and evaluate compounds

Figure 10. Typical stages in the development of a drug. *Reproduced by permission of the Glaxo Group of Companies.*

Under the provisions of the Act, a new medicinal preparation could not be marketed without a Product Licence issued by the minister responsible for health matters. This was valid for five years, and could be renewed. The legislation came into force in September 1971.

The Medicines Commission advises the government of the day how the supply and use of drugs for therapeutic purposes should be controlled. It also suggests which expert committees will be required for this purpose. The best-known of these committees is the Committee on Safety of Medicines, which deals with matters concerning the issue of clinical trial certificates and Product Licences, as well as collecting information about adverse drug reactions. It reports directly to the Medicines Division, the branch of the civil service which advises the appropriate government ministers on pharmaceutical matters. Its professional staff reflect a diversity of interests, including not only administrative civil servants, but also pharmacists, physicians, dentists and lawyers. Amongst them are also to be found the members of the Medicines Inspectorate who frequently visit the premises of pharmaceutical manufacturers. The staff of the Medicines Division also serve the Medicines Commission and the other advisory committees of independent experts, such as the British Pharmacopoeia Commission, the Veterinary Products Committee and the Committee on the Review of Medicines. The latter has been established to decide which of the thousands of products on the market when the Medicines Act came into force should continue to hold their Product Licences that were then automatically granted. It is due to report in 1990. Like the other expert committees, it is not a sub-committee and will thus report directly to the government via the Medicines Division.

Following the pattern established by its predecessor, the Committee on Safety of Medicines has established sub-committees whose members are also experts recruited mainly from academia. The names of these committees adequately describe their functions, *viz.* the Safety, Efficacy, and Adverse Reactions Committee, the Biologicals Committee and the Chemistry, Pharmacy and Standards Committee.

IX

Post-Marketing Surveillance

By the time some drugs have received approval for marketing for the first time, they will have been tested in only a few hundred patients. Others will have been evaluated in, at most, three thousand patients. Unless the incidence of an unexpected adverse reaction is relatively high, it is unlikely that it will have been observed in such a small number of patients. If the particular adverse reaction is a pathological condition that can occur spontaneously, its appearance may not attract attention. The issue that has, therefore, to be considered is whether more thorough clinical evaluation is required before a drug receives approval for marketing.

It is possible to calculate how many patients need to receive a drug with a given incidence of an adverse reaction in order for this to be detected. Thus, if a clearly discernible adverse reaction occurs in 1 patient in every 100, the probability is that it will be detected if 360 patients are treated. Should the incidence be only 1 in 1000, then 3600 patients will be required, and *pro rata* as the incidence becomes lower. For a typical incidence of such a reaction requiring withdrawal of a modern drug from the market, say 1 in 8400 as was apparently the case with benoxaprofen, it can be seen that more than 30 000 patients will need to have received the drug. However, these figures are idealized. Many adverse drug reactions manifest themselves in a manner not dissimilar to normal disease processes which afflict those who are receiving the drug. Obvious examples are jaundice and kidney disease. These will be overlooked much more readily than the phocomelia caused by thalidomide. The calculations then take on a different dimension. Coping with an incidence of only 1 excess occurrence in 100 patients now requires 730 cases to be seen if the background incidence of the reaction being monitored is 1 in a 1000 for the population under consideration. Whilst this is manageable in terms of surveillance, the figure escalates if the incidence of excess occurrences is only 1 in 1000 individuals. This time, 20 000 patients will have to be monitored. Yet such a high incidence for an unanticipated side-effect of a drug that has been thoroughly evaluated before being marketed is exceptional, usually only occurring when a hypersensitivity reaction is involved. Taking a more realistic figure of only 1 excess occurrence of the adverse reaction in 5000 patients, the number requiring to be monitored rises to over 360 000! This is why it seems somewhat unrealistic to imagine that any reasonable increase in the number of patients in pre-marketing clinical trials will make any practical difference to the likelihood of detecting adverse reactions. Literally hundreds of thousands of patients must be exposed to drugs such as those which have been withdrawn from the market in recent years before these will be recognized to pose the level of hazard that they were eventually found to exhibit. If the drug in question is an oral contraceptive with a very low incidence of serious side-effects, and all of which can occur spontaneously, the situation becomes difficult to resolve even when millions of women are taking the drug.

SPONTANEOUS REPORTING OF ADVERSE DRUG REACTIONS

Approaches to the post-marketing surveillance of drugs fall broadly into two categories. The traditional one is drug-oriented, which means that all reports relating to an adverse reaction in those patients receiving a specified drug are examined. This is the method that has always been used within the pharmaceutical industry. Most regulatory authorities now require that companies provide prompt notification of adverse reports received in this way. In some countries (not the United Kingdom) a grey area surrounds the legal position concerning reports which are received by foreign associates of a company. So far as morality is concerned, the position is clear: any company deliberately suppressing information known to it is behaving in a despicable manner.

The alternative approach to drug-oriented monitoring is the disease-oriented one, whereby patients with a specified disease or unusual occurrence are investigated to establish whether it may have been caused by drugs they have been receiving. A striking example is thalidomide, which was incriminated only after a high proportion of the mothers of deformed children were found to have taken the drug. Other examples that can be cited are the detection of the correlation between thromboembolism and the consumption of oral contraceptives, the renal damage caused by phenacetin and other analgesics, and vaginal cancer in daughters of women who had taken stilboestrol during pregnancy. If a large number of patients die as a result of drug toxicity, as in the 1960s when isoprenaline aerosols were associated with an abnormal number of sudden deaths in asthmatics, the distortion of mortality figures may draw attention to the matter. Because such occurrences are, fortunately, rare, current feeling is that it would be worthwhile to introduce a speciality-oriented system of drug surveillance into hospitals. This would focus on the reporting by specialists of pathological conditions that have on previous occasions been associated with adverse drug reactions, e.g. blood dyscrasias, hepatic disease, renal damage, etc.

The most successful system of post-marketing surveillance is probably that which was introduced into the United Kingdom by the Committee on Safety of Drugs in 1963 and which has since been continued by its successor, the Committee on Safety of Medicines. It is familiarly known as the 'yellow card system' on account of the colour of the report card issued to all participants. It relies on the spontaneous reporting of suspected adverse drug reactions by hospital doctors, general practitioners and dentists (surprisingly, pharmacists are not involved). Its effectiveness is thus wholly dependent upon the zeal and competence of those who provide the relevant data to the Committee on Safety of Medicines. Although 12 000 cards are returned annually, it is inevitable that there is a high degree of under-reporting despite strong efforts to improve matters. A study of deaths from thromboembolism amongst women of child-bearing age in the year 1966 revealed that although 53 had been taking oral contraceptives, only eight reports had reached the Committee on Safety of Drugs. Only a quarter of these were from family doctors. The situation has improved since then, but during the period 1972–1980, still only around 16 per cent of eligible practitioners ever sent in reports of adverse drug reactions. Several reasons for this reticence have been advanced, not least of which is complacency engendered by an unsubstantiated belief that the pharmaceutical industry and the Committee on Safety of Medicines will not permit unsafe drugs to be marketed. Unnecessary guilt that a patient has been harmed by a drug supplied in good faith, or possibly even fear of litigation, may deter some doctors from reporting. Others may feel diffident about reporting a reaction because of uncertainty that a drug was responsible. The pressure of work in a busy practice also militates against filling in a yellow card. Ignorance of the nature of adverse drug reactions has been suggested by some critics as a contributory factor to under-reporting.

The yellow card system has supplied sufficient data to identify problems with several drugs before these could be detected by any other investigation. An early example of this was the recognition of the hepatotoxicity of ibufenac within two years of its introduction to the British market in 1966. The drug was then withdrawn by its manufacturer, but not all the adverse reactions detected by the yellow card system require such drastic action to be taken. For instance, revelation of the risk of piroxicam-induced oedema in elderly patients with congestive heart failure was enough to ensure that it could be avoided, thus effectively making this a safer drug. Likewise, the detection of mianserin-induced arthropathy did not necessitate withdrawl of this antidepressant drug which, in many ways, was safer than amitriptyline. Nor did the discovery of rashes caused by protriptyline-induced photosensitivity lead to the removal from the market of this antidepressant noted for its stimulant action in apathetic, withdrawn patients.

The suspension of a Product Licence as a result of an unacceptable incidence of adverse reports has been rare. The first case was that of benoxaprofen, in August 1982. The manufacturer subsequently withdrew the drug, so there was no discussion about lifting the suspension. In 1984, the Product Licence for indoprofen was suspended for three months, whilst that of fenclofenac was not renewed on its expiry.

Any significant difference in the yellow card reports of adverse reactions to a new drug when compared with the therapeutic class profile for similar existing agents is considered by the Committee on Safety of Medicines to be an early warning of a potential hazard. This happened in 1983 with the indomethacin delivery device 'Osmosin', when its ability to cause perforation of the gut indicated a different toxicity profile from that of existing anti-inflammatory drugs.

The ability of the yellow card system to provide a comparison between therapeutically related compounds helps the Committee on Safety of Medicines to reach a decision over what action requires to be taken in the public interest. Withdrawal of a drug from the market may only be called for if the incidence of adverse reactions is seriously out of line with agents of comparable therapeutic value. In fact, it may not always be a new drug which gives cause for concern. As safer drugs become available, the adverse reaction profile of existing ones may be viewed in a new light. This is why 20 years elapsed from the time when phenylbutazone was first viewed with suspicion until decisive action was taken to restrict its use. Similarly, the halving of the maternal mortality rate over the past two decades has upset the equation which formerly balanced the health risk from unplanned pregnancy against that from taking oral contraceptives.

It has never been the intention that the yellow card system should replace skilful observation by the astute clinician. On several occasions, clinical reports published in leading medical journals have drawn attention to problems before there was any heightened awareness of them brought about by the yellow card system. Notable examples are early reports on practolol and benoxaprofen. In both cases, the yellow card system was of considerable value in quickly confirming suspicions that might otherwise have lingered on for many more months. Under its terms of reference, the Committee on Safety of Medicines is required to monitor the literature for adverse effects of recently licensed drugs.

The yellow card system faces drawbacks inherent in any spontaneous reporting of adverse drug reactions. In addition to the under-reporting which has been mentioned, there are problems of both late and selective reporting after attention has been drawn to a potential hazard. For instance, only one report relating to eye damage was received in the first four years during which practolol was on the market, but within weeks of the earliest mention in medical journals that this hazard existed some 200 reports were sent to the Committee on Safety of Medicines. Again, during the first two years in which benoxaprofen was on sale in the United Kingdom, the Committee on Safety of Medicines received reports of 28 deaths attributed to this drug. Shortly after, in May 1982, a paper in the *Lancet* publicly revealed for the first time that benoxaprofen might be a dangerous drug. As a consequence, a further 33 reports of fatalities were submitted to the Committee on Safety of Medicines in the succeeding three months. Even allowing for escalating sales of this heavily promoted drug, the increased rate of reporting is plainly evident.

Compulsory reporting of suspected adverse drug reactions became mandatory in Sweden in 1975. This increased the rate of reporting by only a quarter, so the benefit was minimal. The British system is one of several voluntary ones in use around the world, but it has the highest rate of spontaneous reporting. Each year, British doctors write nearly 400 million prescriptions and return around 12 000 yellow card reports. This level of 262 reports per million of the population is approximately double the rate for the United States scheme, which was introduced in its present form in 1973.

COMPREHENSIVE DRUG SURVEILLANCE

An alternative to voluntary, spontaneous reporting of adverse drug reactions is comprehensive drug surveillance. This requires records to be maintained of all drug exposures and all adverse events, irrespective of whether or not these appear to be linked. The inherent objectivity of this process is ensured by the application of computerized analysis of the frequencies of drug administration and adverse reactions. Such analysis is most likely to detect correlations when the frequency of adverse events is high.

The best-known comprehensive drug surveillance scheme is that which was initiated in July 1966 in three 30-bed medical wards of the Lemuel Shattuck Hospital, Boston. Its success led to expansion of the scheme to include not only many more wards in Boston, but also hospitals in several other countries. The name given to this was the Boston Collaborative Drug Surveillance Program. An attractive feature of the programme was that it did not significantly increase the work load of medical staff in the participating hospitals. A member of the project team, usually a nurse, obtained all the requisite information from the normal clinical records kept in each ward, and interviewed the patient and attending physician to obtain any further information that was required. This included the doctor's graded assessment of the efficacy of each drug. Particular care was taken in recording full details of drug dosage and administration, together with the reasons given for both initiating and terminating drug therapy. When the physician suspected a drug of inducing an adverse effect, a prompt inquiry was carried out by the investigating team. This enabled the programme to provide the information that 0.44 per cent of deaths in hospital could be attributed to drug administration, and that just over a quarter of the patients experienced an adverse reaction to one of the drugs they had been given. A valuable extension of the programme was the recording of drugs taken prior to admission to hospital. This revealed that 6 per cent of admissions were due to adverse drug reactions. Similar schemes to that in Boston have been launched elsewhere, but none has been on such a large scale as the Boston programme.

The unique value of an objective surveillance scheme was shown when the Boston investigators discovered that although 27 out of 105 hospital patients receiving the diuretic ethacrynic acid were noted by their physicians to have developed gastro-intestinal bleeding, none of the cases had been attributed to the drug. Presumably, the doctors believed that this type of event was commonplace. In fact, the incidence of gastro-intestinal bleeding amongst patients who had not received ethacrynic acid was only 4.5 per cent.

MONITORED RELEASE

One way to detect rare adverse drug reactions not seen prior to the marketing of a new drug is to monitor many thousands of patients. In 1910, Paul Ehrlich released 65 000 ampoules of arsphenamine for the treatment of syphilis on the explicit understanding that all investigators who received the drug would furnish details of any adverse reactions. The need for caution in this instance was based on the previous high incidence of serious hypersensitivity reactions to arsenophenylglycine.

The scale on which any surveillance programme of monitored release can be conducted has, in the past, been strictly limited by the immense cost relative to the likely benefit gained by its superiority over spontaneous reporting of adverse reactions. On the other hand, spontaneous reporting relies on each physician recognizing that an adverse reaction might be drug-induced. As has been pointed out, this is much less likely to happen when the adverse reaction is similar to a pathological condition that is usually not drug-related. Clearly, there are advantages and disadvantages with either system of surveillance. The best features of each system were exploited in a form of monitored release known as prescription-event monitoring. This form of surveillance was introduced in the United Kingdom in 1980 by Professor William Inman of the University of Southampton. His programme was financially supported by 28 pharmaceutical companies and had the backing of the General Medical Services Committee of the British Medical Association. In essence, the system relied upon the existence of a central clearing house for National Health Service prescriptions issued by family practitioners in England and Wales. Inman obtained agreement for the two thousand clerks who worked in the Prescription Pricing Authority offices to set aside every prescription containing one of the drugs under investigation. The practical limit was found to be reached with four different drugs at any one time. From the one and a half million prescriptions in the first phase of the programme, every doctor who had prescribed one of the drugs being studied was sent a green form requesting details of any events which had affected not more than four named patients since they had first received the drug. It was explained to the doctors that any incident whatsoever should be recorded, irrespective of whether or not it was thought to be drug-related. Strict confidentiality was guaranteed. The idea was that if a significant number of patients on a certain drug were found, for example, to have had falls then it might indicate unacceptable postural hypotension as a side-effect of that drug. Professor Inman received a very high degree of cooperation from the participating physicians, something in excess of three quarters completing the green forms. Because a strongly promoted new drug can be prescribed more than one hundred thousand times a month in England and Wales, a selection of one sixth of the prescriptions was taken in the initial studies. The results of the study were handled by microcomputers. As experience was gained, the scheme became more efficient. It has now reached the stage where all medicines containing a new drug are surveyed in the first 10 000 to 20 000 patients for whom they are prescribed.

One of the first drugs to be examined by the Southampton Drug Surveillance Unit was benoxaprofen. Only 16 000 patients were subjected to scrutiny, which meant that not enough cases of liver damage to cause concern were detected. Obviously, prescription-event monitoring has the potential to deal with much larger samples, but the question of cost is critical. However, it is still early days so far as this type of surveillance is concerned. Its future scope could be considerable in the light of a recent development in monitored release surveillance.

Interest in monitored release has been revived through the development of modern information technology. Captopril was marketed in the United Kingdom in April 1981 as the first angiotensin converting enzyme inhibitor. Because the early clinical trials had shown it to have the potential to cause serious renal damage or agranulocytosis, the Committee on Safety of Medicines restricted the Product Licence to cover the use of the drug only in those patients with severe hypertension who had not been helped by standard therapy. Since the drug was considered to represent a significant therapeutic development, its manufacturer was particularly keen to have a reliable assessment of its overall safety. In the summer of 1983, the company concerned began to provide nearly one thousand family practitioners with viewdata terminals (consisting of a television set, keyboard and modem) linked via the national Prestel telephone network to a minicomputer. The doctors were then able to key information into the computer each time they examined a patient for whom captopril had been prescribed. This information consisted of medical details, concomitant therapy, and any significant event, irrespective of whether or not it appeared to be drug-related. A unique feature of this surveillance scheme was its interactive nature, through which the participating physicians were able to obtain guidance and information from data stored in the minicomputer. The manufacturer of captopril spent £250 000 on installation costs, but in return expected to receive extensive information on the effects of captopril on twenty thousand patients who would be closely observed for at least one year. The investment proved worthwhile, for the initial results revealed that side-effects were mainly of a minor nature and of relatively low incidence. In October 1985, the licence for captopril was extended to include its use in the treatment of mild to moderate hypertension.

Such was the apparent success of the captopril monitoring scheme that in the winter of 1984 the Department of Health set up its own pilot study into this type of post-marketing surveillance. Four hundred doctors were provided with a viewdata terminal linking them directly to a computer database maintained by the Committee on Safety of Medicines. Two-way communications were again a feature of the system, enabling the participating doctors to receive updated information about adverse drug reactions on request. Using the national Prestel system, the scheme can easily be extended throughout the country if the trial is considered successful.

X

Safer Prescribing

Absolute safety in the administration of drugs is no more attainable than is absolute safety in, say, driving a car. Reducing the risk of adverse drug reactions is the real objective. In principle, this requires balancing the potential benefits of therapy against the known hazards. Let it be said at the outset that despite the obsession of the media with the hazards of drugs, the incidence of serious adverse reactions is low. Over the past 20 years, the Committee on Safety of Medicines on average received reports of just over three hundred drug-associated deaths each year. This represents an incidence of less than one fatality for every million prescriptions that were written.

It is difficult to define what constitutes an adverse drug reaction. Does the term cover inevitable side-effects that have to be accepted as part of a patient's treatment? Certainly, many patients are hospitalized as a consequence of adverse drug reactions; estimates of the proportion have varied to as high as 5 per cent of hospital admissions in the United Kingdom and up to 15 per cent in the United States. This must be balanced against the fact that nearly four hundred million prescriptions are written each year in the United Kingdom, yet only twelve thousand or so reports of adverse drug reactions are sent to the Committee on Safety of Medicines. Even if these returns represented only 10 per cent of the true number of such reactions, the overall incidence would still be less than one in three thousand. Bearing in mind that the majority of the reports describe events that are not unique, many will be due to other factors. When the figures are examined further, it will be seen that they are somewhat distorted by the fact that the group in whom most adverse drug reactions occur, the elderly, account for 30 per cent of National Health Service spending on medicines although they constitute only about 12 per cent of the British population. Many of the drugs given to the elderly are for long courses of treatment, which further increases the risk of adverse reactions. The elderly require many of the drugs which are known to be the most hazardous. The dangers are further compounded by the extent to which the elderly are exposed to multiple drug therapy through the perceived need to palliate a variety of symptoms associated with the ageing process. Such therapy not only multiplies the risks, but also opens up the possibility of troublesome drug interactions.

Since the elderly constitute the group most likely to experience adverse drug reactions, a major step towards safer prescribing has been the identification of the reasons for the problems faced by this group of patients. Many of these problems are caused by reduced rates of glomerular filtration and tubular secretion combining to delay the renal elimination of drugs. Whilst there may be no clinical signs of impairment, the average 80-year-old patient has a glomerular filtration rate (creatinine clearance) of around 50 ml/min, which is approximately half the normal value for younger people. This is comparable with that for a patient with mild renal impairment, so the drugs to be used with caution in such patients are also those for which dosage reduction is advisable in the elderly. For polar compounds which lack sufficient lipophilicity to penetrate into liver cells, elimination by the kidneys is the principal means by which blood levels are reduced. A decrease in glomerular filtration rate causes cumulation of these drugs, and if the therapeutic window is narrow (i.e. there is a low therapeutic index), toxic effects occur.

Given an adequate knowledge of medicinal chemistry, it is sometimes possible to recognize molecular structures which are less likely to undergo extensive metabolism. Generally speaking, drugs that enter the central nervous system do not fall into this category since the lipophilicity required to carry them through the blood–brain barrier ensures their ability to penetrate the liver and undergo metabolism. High lipophilicity also favours passive diffusion of the drug from the glomerular filtrate across the renal tubular wall and back into the circulation. For instance, atenolol and sotalol are beta-adrenoceptor blocking agents with low lipophilicity. This means they are less likely than other beta-adrenoceptor blocking agents to enter the brain and cause sleep disturbance, but they are excreted in the urine largely unchanged. Lithium ion, used in manic depressives, cannot be metabolized and hence must be administered with caution in the elderly and those with renal impairment. Many antibiotics are ionized in the plasma and hence do not readily penetrate liver cells. They are excreted unchanged in the urine in concentrations several times higher than those found in the plasma. Following an intramuscular injection of benzylpenicillin, as much as 90 per cent of the dose appears in the urine within one hour in young adults. This again reflects the chemical nature of the molecule, most of the drug being ionized in the body fluids. One study of patients ranging from 20 to 91 years of age revealed that the half-life of benzylpenicillin varied from 16 minutes to 164 minutes as the glomerular filtration rates in 61 patients varied from 168 ml/min to as low as 9 ml/min. Although most penicillins have a wide therapeutic window and can therefore be used without problems in the presence of renal impairment, this is not so with some of the chemically similar cephalosporins. The dose of these must therefore be adjusted in elderly patients. Aminoglycoside antibiotics furnish further examples of polar molecules that are excreted largely unchanged in the urine. Due to their ototoxic and nephrotoxic properties, these must be used with considerable caution in the elderly and all patients with renal impairment. Other polar antibiotics requiring similar caution include amphotericin, capreomycin, ciclacillin, colistin sulphomethate sodium, cycloserine, tetracyclines (except doxyxycline and minocycline which have higher lipophilicity and thus undergo hepatic metabolism) and vancomycin. Synthetic anti-infective and anticancer agents also present a problem, *viz.* acyclovir, ethambutol, flucytosine, methotrexate and nitrofurantoin. A variety of cardiovascular agents and diuretics require dosage reduction in mild renal impairment, *viz.* bezafibrate, clofibrate, captopril, digoxin, medigoxin, disopyramide, flecainide, procaina-mide and spironolactone. Drugs affecting the endocrine system that should be used with caution include the oral hypoglycaemics acetohexamide, chlorpropa-mide and metformin, as well as the antithyroid drug propylthiouracil.

When prescribing a drug that will accumulate if the glomerular filtration rate is low, the dosage quantity is reduced, initially to around half the usual adult amount. Subsequently, the dosage can be adjusted in the light of the clinical response. An alternative approach is to reduce the frequency of dosage or else to select an analogue that is known to undergo extensive metabolism. For instance, the methyl group in tolbutamide is rapidly oxidized by microsomal enzymes in the liver. This drug is much less likely to cause hypoglycaemia in the elderly than would chlorpropamide in which there is a chlorine atom in place of the methyl group.

The second consideration affecting drug action in the elderly is that plasma levels of some drugs which would be safe in younger patients can be hazardous; this is presumed to be due to heightened sensitivity of either the target organ or other tissues. Drugs acting on the central nervous system are striking examples of this, with hypnotics, antidepressants, mild tranquillizers (such as the benzo-diazepines) and major tranquillizers (tricyclics, butyrophenones), and anti-parkinsonian drugs producing disproportionate responses. There is also evi-dence to suggest that anticoagulants such as warfarin, which inhibits vitamin K synthesis, also produce a stronger reaction in elderly patients.

A third factor causing drug-related problems in the elderly is inappropriate prescribing. The obvious examples of this relate to the excessive use of hypotensive agents and diuretics. A fall caused by postural hypotension induced through over-zealous prescribing could be a greater hazard than that presented by mild hypertension itself. As for diuretics, apart from causing distressing incontinence in some patients, they can bring about potassium depletion or dehydration, either of which may pass unnoticed in an old person.

Poor compliance with the wishes of the prescriber is a common problem with elderly patients. It may simply be due to inability to interpret the prescriber's instructions because of deficiencies of hearing or eyesight. Sometimes it is the presentation of the product that is at fault, as when an elderly patient is inadvertently supplied with tablets in a child-resistant container, or tablets are either too small to be counted out or too large to be swallowed conveniently. Often, lack of compliance is caused by the practice of prescribing several drugs at a time in an effort to control a variety of symptoms of the ageing process. Inevitably, the patient becomes confused by the different times of administra-tion for each type of medication. When different types of tablets are taken simultaneously the patient may forget which one has already been swallowed. Multiple prescribing is also inadvisable because it opens up the possibility of harmful drug interactions. These are not limited to direct pharmacological incompatibility, but can also arise through interference with absorption, distribution, metabolism or excretion of other drugs. For example, the anticho-linergic activity associated with many compounds will delay gastric emptying and alter intestinal motility sufficiently to change the absorption characteristics of many other drugs. This can be hazardous when administering compounds with a narrow therapeutic window, e.g. digoxin, phenytoin or lithium. Another way that the transport and distribution of one drug may be influenced by another is seen in the competition between acidic drugs for binding sites in plasma albumin. Displacement of a drug with a narrow therapeutic window from its inert protein-bound complex can produce toxic levels of the free drug. This could be lethal in the case of warfarin or methotrexate, these being capable of displacement by aspirin and other non-steroidal anti-inflammatory agents, sulphonamides, sulphonylurea oral hypoglycaemics or anticonvulsants. Other possible interactions with acidic drugs can arise from competition for saturable active transport excretory processes in the proximal renal tubules. Acidic compounds known to be actively secreted in this manner include penicillins, sulphonamides, non-steroidal anti-inflammatory agents, uricosuric agents and thiazide diuretics.

Drugs may interact even when not simultaneously administered. A marked increase in the activity of liver enzymes can be induced by the chronic administration of a small number of drugs, notably barbiturates, carbamazepine, phenytoin and rifampicin. When other drugs are administered to patients receiving such agents, due allowance must be made for this. For instance, a patient receiving one of these drugs may be stabilized on warfarin, but on withdrawal of the enzyme-inducing agent, toxic levels of the anticoagulant will arise unless dosage adjustment is made to compensate for reduced metabolic detoxification. Occasional reports have appeared claiming that female patients receiving phenytoin, phenobarbitone or rifampicin may experience oral contraceptive failure unless dosage is adjusted. Fears have been expressed that similar contraceptive failure can occur if antibiotic therapy with broad-spectrum agents such as ampicillin results in destruction of the intestinal bacteria which break down steroid conjugates recycled by the entero-hepatic circulation. The action of such bacteria is said to ensure reabsorption of the free steroid.

The problems faced by the elderly typify the commonest types of prescribing hazard, namely unintentional overdosing and drug interactions. One of these problems particularly applies to infants in the first month of life, namely that caused by low renal clearance of drugs, but in their case it is coupled with immature hepatic metabolism. At birth, the glomerular filtration rate in the full-term infant is around 2–4 ml/min, increasing to 8–20 ml/min by the third day, followed by a gradual increase until normal values are attained in the second 6 months of life. The impairment of tubular secretion is even more marked. The effect these factors have on blood levels of drugs is complicated by several others. Although most microsomal enzymes are present at birth, their level of activity is extremely low, only increasing gradually during the first few weeks of life. This means that drugs which in adults are extensively metabolized have instead to rely on inadequate renal elimination. The influence of the higher proportion of body water (75–85 per cent) in new-born infants compared to adults (50–55 per cent), due to a greater volume of extracellular fluid, further distorts dosage calculations when administering drugs. So, too, does the diminished plasma protein binding capacity. In addition, it should be kept in mind that the gastro-intestinal tract of the neonate represents a larger proportion of the body than in later life, hence drug absorption patterns are altered. At the same time the rate of gastric emptying may be prolonged in any infant under the age of six months, resulting in a greatly delayed onset of drug action. Absorption from the upper reaches of the small intestine can also be delayed because of irregular peristaltic movements. Taking all these circumstances into account, it becomes apparent why dosage calculations based on age have proved most unreliable. Weight or, preferably, body surface area calculations are a better guide, but caution is still necessary in assessing dosage in very young infants. The situation is even more hazardous in the low birth weight infant. Since the commonest drugs required to be administered to young infants are antibiotics, it is always desirable to select one with a wide therapeutic window whenever possible.

All patients, but especially the elderly, face problems arising from the intrinsic pharmacological properties of many drugs, particularly those acting on the autonomic nervous system. Most drugs designed to influence neurotransmission lack selectivity and produce unwanted actions throughout the body. It is well-recognized that anticholinergic agents do not merely reduce gastric acid secretion or relieve spasm, but also cause dry mouth, blurred vision, tachycardia, constipation and urinary retention, especially in the presence of an enlarged prostate. However, it is sometimes forgotten that strong anticholinergic activity is present in drugs such as imipramine, amitriptyline and other tricyclic antidepressants, and antihistamines (including those prescribed for prevention of travel sickness), as well as benzhexol and orphenadrine used in parkinsonism. Similarly, agents used to interfere with sympathetic activity will create a variety of problems. For instance, beta-adrenoceptor blocking agents can produce bradycardia or bronchoconstriction, a particular hazard for asthmatics, and also interfere with the response of diabetics to hypoglycaemia. Adrenergic neurone blocking agents such as bethanidine, debrisoquine and guanethidine inevitably induce postural hypotension and can prevent ejaculation. The same side-effects can appear with chlorpromazine, which was originally designed to reduce sympathetic activity by its central action. Most other major tranquillizers exhibit similar side-effects. A further problem is presented by the ability of tranquillizers, particularly those butyrophenones and tricyclics containing a piperazine ring, to induce a variety of extra-pyramidal effects through their ability to block dopaminergic receptors in the central nervous system. These include parkinson-like tremor, a characteristic form of restlessness known as akathisia, dystonia of certain muscles in the head, and tardive dyskinesia. For elderly patients living in poor housing conditions in cold climates, the ability of phenothiazines, benzodiazepines and barbiturates to induce hypothermia is a disturbing phenomenon.

The other main group of hazards is faced by all patients, namely hyper-sensitivity reactions. These are largely unpredictable, although sometimes avoidable by means of sensitivity testing or recognition of a history of allergic responses. For established drugs that have been on the market for many years, the risk has usually been quantified and will be taken into account by the prescriber when making a risk-benefit assessment of the proposed course of treament. With any drug that has only been introduced within the previous two or three years, it is prudent to assume that this type of hazard exists until evidence to the contrary becomes available.

XI

Drugs and the Public

It is universally accepted that informed consent must be received from all patients participating in a clinical trial. Whether this view is considered from the perspective of the self-interest of the physician wishing to cover himself against criticism and litigation should any mishap arise, or from that of acceptance of the basic right of the patient to be informed of any risk being faced, or even from that of a belief that the decision to participate in a trial should be a shared one, does not matter. The ethical position is unambiguous. Why, then, does the need for informed consent not also apply when a patient is exposed to the greater hazard of being given a medicine without as thorough an examination before receiving it, and without the monitoring and after-care that always accompany administration of a new drug undergoing clinical trial? There is, after all, little evidence to show that more adverse reactions occur during clinical trials than after a medicine is routinely prescribed by a physician. So far as elderly patients, at least, are concerned, it would not be surprising if the evidence were to show that they would be safer participating in a clinical trial than taking a drug prescribed by their general practitioner. Lest this admittedly contentious statement be misconstrued, let it be said at once that no criticism of the practitioner is implied. The reality of the situation is that he cannot routinely afford to provide the facilities and ancillary support which large pharmaceutical companies pay for when trials of their new products are organized. This could change if society accepted that drugs should be prescribed far less frequently and with thorough monitoring of response by laboratory and clinical methods. The already over-stressed family practitioner would find it difficult to provide the necessary facilities, but the modern pharmacy graduate can offer valuable assistance if given the resources. Meanwhile, the optimist will wait patiently for this change eventually to occur in the twenty-first century!

At present, patients are not altogether uninformed about the medicines they receive. Some of their information comes from the doctor at the time of prescribing, and some from the pharmacist when the medicine is dispensed. The introduction of computerized labelling systems enables pharmacists to complement their verbal advice with essential printed information relating to administration of the prescribed medicine. In Sweden, pharmacists routinely supply a leaflet about the class of drugs that has been prescribed. Information is also provided by manufacturers who include package inserts for the patient, but the role of these in helping to improve patient compliance with the prescriber's wishes has been questioned. The increasing trend towards original pack dispensing certainly makes it easier for doctors to ask pharmacists to ensure that the inserts are provided. In the United States, ever-increasing amounts of information are being crammed into package inserts by manufacturers who are anxious to ensure that they cannot be sued in the courts on the grounds that the patient had not been informed that a specific rare side-effect might occur. This has not happened in the United Kingdom, where the wording of package inserts is governed by the Medicines (Leaflets) Regulations 1977. However, doctors who rely on a well-written package insert being included with a particular product should be aware that if either another manufacturer's brand of the same drug or a generic equivalent is dispensed, the patient will not receive the intended leaflet.

Whether desirable or not, an increasing amount of information about medicines is disseminated by the media. Since the thalidomide disaster, much of the media coverage of drugs has focused on adverse reactions rather than on the benefits of medicines or how to use them wisely. Disasters are always more newsworthy than are medicines which achieve modest improvements in mankind's lot without creating difficulties.

There are over a thousand medicinal compounds, and several times as many formulations of them, on the market. It is thus inevitable that the commercial success or failure of any new product will be strongly influenced by the amount of early publicity it receives, irrespective of whether that publicity emanates from independent sources or from the drug company itself. Vast sums of money are spent by the pharmaceutical industry on promotion of new drugs in order to recoup the millions of pounds invested over many years (it could now be as much as £100 million) without any financial return until the drug reaches the market. If a new drug proves unsafe, it will be withdrawn unless there is good reason for prescribing it for particular groups of patients who are either unlikely to be harmed, or else stand to gain so much benefit that the hazard is worth confronting. Unfortunately, bad publicity in the media can result in the unnecessary withdrawal of a useful drug. There are those who feel this has happened recently as a consequence of the media and the general public alike being unable to recognize that no drug will ever be entirely safe.

It should be evident from what has been written in the preceding pages that, when a new drug is launched, the question of its safety has still to be resolved. The first two or three years after a drug is newly marketed will witness many reports of adverse reactions, both trivial and severe. Fortunately, the risk-benefit ratio for most new drugs turns out to be favourable. Where this is not so, unless prompt action is taken by either the manufacturer or a regulatory authority, the public will be dramatically informed of this by the ever-vigilant media. There can be no quibble over that, but situations do arise where the issue is not clear-cut and the decision whether or not to retain the drug on the market is debatable. When the debate takes place amongst professionals who are fully acquainted with all the factors involved, a satisfactory conclusion is usually reached fairly quickly. Occasionally, the debate enters the public arena, where not all the participants are qualified to pass judgement on the pharmacological, toxicological and therapeutic properties of the drug concerned. Many journalists and broadcasters are, nevertheless, eminently well-qualified to probe the frailty of the human condition if they suspect commercial interests and professional indifference have acted against the welfare of the sick and infirm. The debate will certainly receive publicity if journalists feel that sufficient newsworthy material is involved. This, of course, depends on the attitude of the new medium rather than on what the medical profession or the pharmaceutical industry may perceive to be the merits or otherwise of the case. After all, it is the remit of every journalist to tell his story, and there will be more human interest inherent in some matters than in others. A feature describing the death of a patient from a dental anaesthetic is more likely to appear than one about a cancer patient who dies after receiving an experimental drug. The reason is obvious: most people realize they might require a dental anaesthetic, and they expect it to be safe. Similarly, the widespread use of oral contraceptives ensured that maximum media coverage would be given to the findings of the major inquiries into their safety. In this particular case, the experts are themselves perplexed. This encourages journalists to venture forth where angels fear to tread!

An important factor determining the likelihood of coverage is the perception by the media of the risks versus the benefits that are involved when medicines are prescribed. If any of the participants in the public debate about the safety of a medicinal product fail to indicate the true reality of the risks it presents, as opposed to its benefits, the resulting sense of outrage by the opposing faction will be justified. The medical profession describes as unbalanced reporting that which presents only one side of the equation, be it uncritical enthusiasm for an unproven new drug for which there is scant evidence of adequate safety, or the opprobrium heaped upon the manufacturer of a drug that is held to be unacceptably dangerous despite its proven ability to save lives that could otherwise have probably been lost. Particular anger is expressed by the medical profession when a crusading (or possibly malicious) journalist or broadcaster fails to present the facts necessary for making a balanced judgement. Such anger is matched by that understandably felt by journalists who are told they cannot be expected to pass such judgement.

There can be no criticism of the German newspaper that hurriedly sensationalized the thalidomide disaster and thereby accelerated the withdrawal of this drug from the market six days after Dr Lenz first suggested it might be teratogenic. The newspaper rightly considered the risk of deformity to be out of all proportion to the benefits likely to have been conferred by a mild sedative. The bitter controversies that do arise are mainly those which concern drugs where the issues are not black and white, but rather are in shades of grey. The inability to quantify what is an acceptable risk lies at the heart of most of these. How can the improvement in the quality of life offered to a crippled octogenarian by a new antiarthritic drug be measured? And how can this be weighed against the slight possibility of killing that patient when that risk can be reduced by skilled administration of the drug? These are the questions raised by the recent controversy surrounding benoxaprofen.

Doctors, pharmacists and biomedical scientists are, on occasion, asked their opinion about a press report of a new drug that has just been discovered. Sometimes, the questioner is desperately ill, or else is a close relative of such a person. To answer by explaining that it may be ten years or more before the new discovery can be harnessed to furnish a medicine may well shatter the hopes of the inquirer, but all too often that is the reality. There are many substances with interesting pharmacological activity, but all too few of these are ever developed into safe and effective medicines. Unfortunately, the pharmacological revolution that followed in the wake of the development of penicillin in the 1940s created an uncritical enthusiasm for new drugs by the public. This was abruptly tempered by the thalidomide disaster. Since then, there has been a paradoxical blend of the fear that another catastrophe lies around the corner, coupled with a fervent expectation that more miracle drugs are about to be discovered. Naturally, the media quickly recognized this. There have been as many newspaper articles and broadcasts which sang the praises of an unproven new product as there have been exposures of 'drug scandals'. The former are of lesser consequence than the latter. With a healthy suspicion of large organizations, vested interests and even the professions, journalists have not been slow to seize on any hint that all may not be well with a new drug. Sometimes they fail to give a balanced presentation simply because they have been unable to acquire all of the necessary information in time to meet a predetermined production deadline. They protest in their own defence that intense competition from rival newspapers or broadcasting stations forces them to rush into print or on to the air with exclusive coverage before they have time to ferret out all the facts. Where would we be if such a feeble excuse were offered by the pharmaceutical industry to justify the premature release of a drug on to the market?

Journalists are understandably aggrieved when they cannot elicit the necessary information about a problematic drug from participants in a radio or television programme. There can be various reasons for this, not least of which is the reluctance of pharmaceutical companies, regulatory authorities or doctors to disclose prematurely any facts which they believe will be misrepresented and thus cause unwarranted public alarm before their investigations have been completed. But how does a journalist differentiate between this and what he would describe as 'a cover-up'? At the heart of the problem is at best impatience, and at worst a lack of trust between the parties concerned. The journalist or interviewer may feel the person being questioned is seeking to hide the truth, whilst at the same time his own background knowledge of the subject in question is being doubted by his adversary. Where this mutual distrust can be broken down by prior discussion and a frank exchange of views, the likelihood of a balanced presentation is increased.

Some readers of this book may feel that the views of the media are of little consequence, especially those of the popular press and television. This attitude is mistaken. The man in the street knows little of the Food and Drug Administration or the Committee on Safety of Medicines, but he does read a newspaper and watches too much television. Rightly or wrongly, he has come to accept the press as a watchdog over the affairs of the pharmaceutical industry. When a drug is condemned by the media, the attitude of any patient receiving it will possibly be distrustful. Such bias would be unacceptable in a properly controlled clinical trial, so one must accept that it would be equally undesirable in other situations. Knowing this, a reputable manufacturer may feel forced to withdraw a product from the market.

The cost of bringing a new drug to market continues to escalate. Protracted boardroom discussions are held to decide whether it can be worthwhile gambling with shareholders' money by investing it in the future of a new compound whose value cannot be fully assessed until it has reached the market. Any factor which reduces the chances of success in the marketplace will weigh against a decision to proceed. Drugs already face trial by the medical profession and regulatory authorities. If, in addition, they are to be judged by the media, the possibility of unfair condemnation becomes a worrying complication that may count against the decision to proceed with a new drug.

It should be evident from what has just been said that there are times when it is desirable that the public be informed by the media about adverse effects of drugs. The process that is involved has to be carefully considered. If a drug poses a threat to the health of the community, then a speedy disclosure is essential. However, failure to ensure that doctors and pharmacists are informed as quickly as the public will mean that any individual seeking professional advice may find it is not forthcoming. This can add to the distress that may already have been experienced. One way round this problem has been tried successfully in the United Kingdom. This involves the regulatory authority providing a full disclosure for the media on the understanding that none of the information will be released until an agreed time. The period of the embargo must be kept to the absolute minimum necessary to ensure that the medical and pharmaceutical professions receive all essential information. With modern technology, this can be as little as two or three days.

It is reassuring to note how few drugs have to be withdrawn on grounds of safety. Between 1960 and 1982, only 14 drugs were removed from either the United Kingdom or the United States markets specifically on these grounds. There has certainly been a spate of withdrawals of antirheumatic agents in the past year or two, occasioned to a large extent by the benoxaprofen incident. The existence of safer alternatives was, of course, revealed by modern surveillance techniques. In the past, it was easy to be critical of the pharmaceutical industry for introducing 'me too' drugs which had activities similar to those of existing products. Such developments now have to be viewed in a different light. The future will continue to witness a steady stream of drugs being withdrawn as they outlive their usefulness when compared with new products adjudged to be safer and just as efficacious.

In the quarter of a century since the thalidomide disaster, much progress has been made towards increasing the safety of medicines. The material price, whilst surely justified, has been high. The fear is that a combination of colossal developmental costs and limited patent coverage remaining after the long period of gestation of new medicines will combine to discourage the development of drugs capable of alleviating less common diseases and those which afflict the inhabitants of the Third World. It would seem that the patent situation could be remedied at the stroke of a pen by the simple expedient of providing coverage from the date of first marketing. Simple, that is, except for the legislators who would face a minefield of difficulties in establishing new principles to govern patent law when applied specifically to pharmaceuticals. The problem of development costs is even more complex. Public agencies support basic research into rare or tropical diseases, but for them to lay out tens of millions of pounds to cover developmental costs with no guarantee that a safe product will emerge at the end of the day is quite another. It might be thought that one possible solution could lie in agreements with industry to look at the overall profits of each company with a view to ensuring that a proportion of these is invested in the research and development of what have been called orphan drugs. This solution could not be introduced unilaterally since pharmaceutical companies operate on an international scale and profits in any one country can be notional insofar as they depend on company purchases of materials and equipment from, and sales to, other countries. An internationally agreed levy on profits is required, the proceeds from which would be reinvested in research and development of orphan drugs through the issuing of contracts to the pharmaceutical industry. Those companies who already conduct such investigations should be the first to be awarded contracts.

Further Reading

CHAPTER 1

Albert, A. *Selective Toxicity: Physico-chemical Basis of Therapeutics*, 7th edn, Chapman & Hall, London, 1985.
Sneader, W. *Drug Discovery: The Evolution of Modern Medicines*, John Wiley, Chichester, 1985.

CHAPTER 2

Fishburn, A. *An Introduction to Pharmaceutical Formulation*, Pergamon, Oxford, 1965.
Ganderton, D. Dosage forms of the present and future, *Pharm. J.*, 1984, *234*, 349–351.
Gennaro, A.R. (ed.) *Remington's Pharmaceutical Sciences*, Mack Publishing Co., Easton, Pa, 1985.
Lees, K.A. Pharmaceutical product development—the management problem, *Pharm. J.*, 1972, *222*, 255–258.

CHAPTER 3

Lasagna, L. The pharmaceutical revolution: its impact on science and society, *Science*, 1969, *166*, 1227–1233.
Stenlake, J.B. Medicines today: quality and safety, *Pharm. J.*, 1984, *234*, 346–349.
Thomas, W.G. Quality assurance and the patient, *Pharm. J.*, 1978, *228*, 235–237.

CHAPTER 4

D'Arcy, P.F. and Griffin, J.P. (eds) *Iatrogenic Diseases*, 3rd edn, Oxford University Press, Oxford, 1986.
Davies, D.M. (ed.) *Textbook of Adverse Drug Reactions*, 3rd edn, Oxford University Press, Oxford, 1986.
Gorrod, J.W. (ed.) *Drug Toxicity*, Taylor & Francis, London, 1979.
Reynolds, J.E.F. (ed.) *Martindale. The Extra Pharmacopoeia*, 28th edn, The Pharmaceutical Press, London, 1983.

CHAPTER 5

De Weck, A.L. and Bundgaard, H. (eds) *Allergic Reactions to Drugs*, Springer-Verlag, New York, 1983.

CHAPTER 6

Balls, M., Riddell, R.J. and Worden, A.N. (eds) *Animals and Alternatives in Toxicity Testing*, Academic Press, London, 1983.
Gorrod, J.W. (ed.) *Testing for Toxicity*, Taylor & Francis, London, 1981.
Hayes, A.W. (ed.) *Principles and Methods of Toxicology*, Raven Press, New York, 1982.
Laurence, D.R., McLean, A.E. and Weatherall, M. (eds) *Safety Testing of New Drugs. Laboratory Predictions and Clinical Performance*, Academic Press, London, 1984.
Triana, V.M. The role of toxicology in drug research and development, *Medicinal Research Reviews*, 1983, *3*, 43–72.

CHAPTER 7

Pocock, S.J. *Clinical Trials: A Practical Approach*, John Wiley, Chichester, 1983.
Waife, S.O. and Shapiro, A.P. (eds) *The Clinical Evaluation of New Drugs*, Hoeber-Harper, New York, 1959.

CHAPTER 8

Hamner, C.E. (ed.) *Drug Development*, CRC Press, Boca Raton, Fa, 1982.

CHAPTER 9

Inman, W.H. (ed.) *Monitoring for Drug Safety*, MTP Press, Lancaster, 1980.
Jick, H., Miettinen, O.S., Shapiro, S., Lewis, G.P., Siskind, V. and Slone, D. Comprehensive drug surveillance, *J. Am. Med. Ass.*, 1970, *213*, 1455–1460.

CHAPTER 10

World Health Organization. Health care in the elderly: report of the technical group on use of medicaments by the elderly. *Drugs*, 1981, *22*, 279–294.

Index